International Conference on Human Factors 2018

London, United Kingdom
26-27 September 2018

ISBN: 978-1-5108-8320-8

Printed from e-media with permission by:

Curran Associates, Inc.
57 Morehouse Lane
Red Hook, NY 12571

CURRAN ASSOCIATES INC. proceedings.com

Some format issues inherent in the e-media version may also appear in this print version.

Printed by Curran Associates, Inc. (2018)

For permission requests, please contact The Royal Institution of Naval Architects
at the address below.

The Royal Institution of Naval Architects
8-9 Northumberland Street
London, WC2N 5DA
United Kingdom

Phone: 44 207 235 4622
Fax: 44 207 259 5912

publications@rina.org.uk

Additional copies of this publication are available from:

Curran Associates, Inc.
57 Morehouse Lane
Red Hook, NY 12571 USA
Phone: 845-758-0400
Fax: 845-758-2633
Email: curran@proceedings.com
Web: www.proceedings.com

CONTENTS

Human Factors

26 - 27 September 2018, London, UK

PAPERS

THE ROYAL INSTITUTION OF NAVAL ARCHITECTS
8-9 Northumberland Street, London, WC2N 5DA, Telephone: +44 (0)20 7235 4622

The Royal Institution of
Naval Architects

INTERNATIONAL CONFERENCE

Human Factors

26 - 27 September 2018, London, UK

PRESENTATION

THE HUMAN FACTOR IN THE DIGITAL AGE

M Lundh, K Aylward, Y Man and S MacKinnon, Chalmers University of Technology, SWEDEN

SUMMARY

The shipping industry is traditional and conservative, with slow-moving regulatory body acting in a reactive manner often driven by major shipping disasters. However, moving into the digital age in the 4th IR does not exclude the shipping industry which now is facing a development towards smart and highly automated vessels driven by rapid technological developments. A fully autonomous electric powered container vessel can be a reality in 4 years.

The transition into highly and fully autonomous vessels will change the seafarer as we now know them. No doubt, the traditional core knowledge described in the STCW Convention will still be valid for the future operators. The question is what else to they need to know and what skill sets do they need to have? This paper aims at discussing the challenges involved in preparing the future seafarers in their coming role in this digital age.

1. INTRODUCTION

In the middle of the eighteenth century, the first industrial revolution (IR) took off. The invention of the steam engine started the development of industries changing leading to a transition from farming feudal society to manufacturing industries [1, 2]. This was followed by the second and third industrial revolution moving society from a physical labour force into mental efforts [2]. These phases in industrial development, all created major societal changes and opportunities which have made major impacts on how we work and live [3]. Partly overlapping the technical advancements made during the 3rd IR we now are facing the challenges in the 4th IR. It argued that this revolution is different from the previous ones due to "...emerging technologies and broad-based innovation are diffusing much faster and more widely than in previous ones..." and is different in scale, scope and complexity characterized by a fusion of technologies" [3, page 8]. The shipping industry on the other hand, is traditional and conservative, with a slow-moving regulatory body acting in a reactive manner often driven by major shipping catastrophises [4] However, the 4th IR is not excluding the shipping industry which now is facing a development towards smart and highly automated vessels driven of a rapid technology development. The aim of this transformation is a safer and more efficient shipping industry [1, 5]. The development towards a highly automated operation and monitoring of vessels creates a wide range of challenges e.g. technological, operational and monitoring of vessels, cyber security, regulatory and legal challenges [1]. This paper aims at discussing the challenges involved in preparing the future seafarers in their coming role in this digital age.

2. AUTOMATION AND THE DIGITAL AGE IN SHIPPING

Brainbridge's paper "Ironies of Automation", is still often cited in scientific articles discussing the impact of automation [6]. Her concluding sentence in this article "...the irony that one is not by automating necessarily removing the difficulties, and also the possibility that resolving them will require even greater technological ingenuity than does classic automation..." is more valid than ever. When reading industry's response to the challenges ahead created by this digital age, quotations like "redefining shipping" and "transition" become more common. [7, 8]. Areas of discussions are technology, cyber security, regulations and legal challenges together with operational and management of a future fleet port development and logistic [9].

The shipping industry has been the subject of technical changes for several decades [10, 11, 12]. The technological developments that the shipping industry has been subjected too and its implications on crew performance are well researched [10, 13, 14]. Among the results we see evidence of a change in task performance, e.g. new tasks to perform, traditional tasks performed in different ways and a demand for a different vocational skill sets and knowledge. The ships' bridges and the introduction of new navigational tools is often used as an example to illustrate this development [10, 14]. Figure 1 illustrates how navigational aids have increased over the years [14]. The purpose for this development has been to increase safety and efficiency, but research has suggested that these technologies can be both counterproductive and ineffective [10].

Recently finished and ongoing EU funded research are trying to understand how the introduction of digital tools affect the operators´ performance their relationship to safety and efficiency in the operation of vessels (Sea Traffic Management (STM), MonaLisa 2.0, EfficienSea and EfficienSea II) [15, 16 17]. STM aims at "taking maritime transport into the digital age" [15]. It focuses on creating a safer, more efficient and environmentally friendly maritime sector. The goals for full deployment of STM resources by the year 2030 is a 50% reduction of accidents, a 10% reduction in voyage costs and 30% reduction in waiting time for berthing, 7% lower fuel

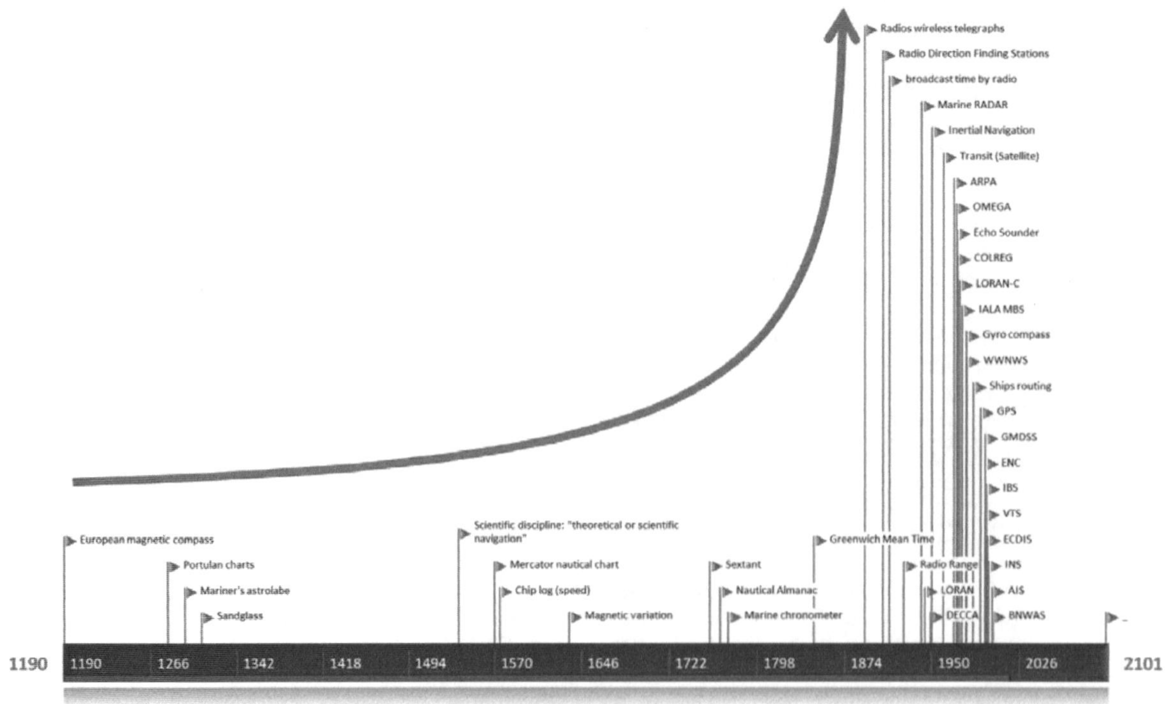

Figure 1 The development of navigational aids [14] (With permission)

consumption and 7% lower greenhouse gas emissions compared to 2015 values.

In 2012 the Maritime Unmanned Navigation through Intelligence in Networks (MUNIN) started [18]. MUNIN was a part of the 7th framework EU-projects and the idea behind it was to develop and verify a concept for an autonomous ship, which is defined as a vessel primarily guided by automated on-board decision systems but controlled by a remote operator in a shore side control station. The European Defense Agency project Safety and Regulations for European Unmanned Maritime Systems (SARUMS), focused on legal and liability aspects [19].

Following the first discussion on autonomous vessels was the introduction of the concept of smart vessels [8]. At the end of 2017, the China Daily reported that the world´s first smart ship had started its trial voyage [20]. The ship was equipped with a marine system with an autonomous learning ability and intelligent operation system. The system is said to analyse real time navigation and meteorological data, pick the best route and alert the crew to hidden dangers in advance. The technical performance of the ships has passed the official assessments from the China Classification Society and Lloyd's Register of Shipping.

A press release in June 2018 revealed a cooperation between Stena and Hitachi introducing Artificial Intelligence (AI) into the picture calling it "Cognitive Shipping" [21]. Yet another press release in August 2018 announced that Samsung Heavy industries was turning to Amazon Web Services, one of the world's leading cloud provider, to expand their smart shipping capabilities using the breadth and depth of AWS's services, including machine learning, augmented reality and virtual reality, analytics, databases, compute, and storage [22]. August 15th 2018 YARA announced that in partnership with the technology company Kongsberg that the Norwegian shipbuilder VARD are to build the first autonomous, electric powered container vessel, Yara Birkaland. The vessel is to be launched in 2020, gradually move from manned operations to fully autonomous operation by 2022 [23].

3. REGULATORY AND CLASS RESPONSES ON DEVELOPMENTS IN THE DIGITAL AGE

A report addressing the results in MUNIN and SARUM concludes "both projects concur with that we can't wait for IMO for a world ratification, since national initiatives allow autonomous vessels on their domestic waters. We will have parallel development of autonomous systems and the technology will mature faster than the legislation/liability will be finalized within IMO" [24].

The shipping industries' transition into the digital age has a lot of regulatory challenges as the existing maritime rules and regulations as all assume by default that a ship is manned. However, IMO took its first step towards addressing the challenges in autonomous vessels in their

Maritime Safety Committee's 99th session on the 16-25 May 2018, at IMO Headquarters in London (25). The purpose of this step was to investigate the operation of Maritime Autonomous Surface Ships (MASS) with regards to safety, security and environmental impact. To facilitate this approach, IMO organizes the levels of autonomy in four categories;

- Ships with automated process and decision support (seafarers on board but some operations are automated)
- Remotely controlled ships with seafarers on board (Small crew onboard, ship is controlled and operated from another location)
- Remotely controlled ship without seafarer on board (ship is controlled and operated from another location)
- Fully autonomous ship (the operating system is supposed to be able to make decisions and determine actions by itself)

IMO notes that during a voyage, different levels of automation can be used. The work of addressing MASS must be undertaken together with the legal challenges, the human element, technology and operational factors [25, 26].

In addition to IMOs definition and to provide some clarity for ship designers Lloyd's register has defined 6 levels of automation [27]. These two ways of defining demonstrate similarities;

- AL 0 Manual – No autonomous function
- AL 1 – On-ship decision support
- AL 2 – On and off-ship decision support
- AL 3 – "Active" human in the loop
- AL4 – Human in the loop – operator/supervisory
- AL 5 – Fully autonomous (but with rarely supervised operation)
- AL 6 – Fully autonomous

Bureau Veritas has followed with "Guidelines for Autonomous Shipping" [28]. Their guidelines also define levels of autonomy ranging from 0 to 5; Human operated, directed, delegated, supervised over to fully automated.

Concern has been raised about the lack of regulatory support in the area of automatization and digitalization [24]. The classification societies are however putting much effort into keeping up with the developments of the industry. In a press release earlier, the International Association of Classification Societies (IACS) stated "In an industry undergoing rapid change, the IACS Chairman identified the need for classification societies and IACS itself to be adaptable and prepared for change, while staying true to the core purpose of classification" [29].

The pictures presented by these bodies are complex. It covers different levels of automation, even emphasizing that different levels of automation can be used on board depending upon operation mode [25, 26]. Today's regulatory support is based on manned vessels as the regulatory bodies never really envisioned unmanned vessels. Clearly industry, regulatory bodies and the IMO are preparing to meet the future challenges and making qualified estimations on how it is going to look like and what regulatory support that is going to be needed. Rod Johnson, director of marine risk management and a member of the committee of the International Union of Maritime Insurers, was interviewed on the topic of questions raised with regards to crewless ships [44]. Once autonomous ships are a reality, "the entire current legal framework for maritime law and insurance is done," said Johnson. "The master has not been replaced; he is just gone. Commodity ships (bulk carriers) would be most amenable to that technology. I'm not overly bothered by fully automated ships, but I am extremely bothered by heavily automated ones."

4. VOCATIONAL SKILL SETS IN SHIP OPERATIONS IN THE 4TH INDUSTRIAL REVOLUTION

Johnson's statement above illustrates well the situation ahead of us. Although autonomous vessels can be a reality in four years' time [24] we will still have older, more traditional vessels sharing the seas with highly automated and autonomous vessels.

The topic of skills necessary in the 4th IR is regularly discussed and research in general terms of competencies necessary in the 4th IR is ongoing [30, 36]. Xu et al [1] emphasized the speed and measures of the changes coming in the 4th IR [1]. They foresee advantages e.g. a more active role for Artificial Intelligence (AI), a fusion between technology and domains and the connected life through internet. They also anticipate challenges like automation substitute for labour across the entire economy and that low skilled jobs will be replaced by computers and digitalization. Schwab [3] thinks that skills and talent will be a critical factor while Blinder [31, page 16] argue that "the nature of education should evolve along with the nature of skills demanded of the workforce" in the 3rd IR and that "education should be designed with the nature of future jobs in mind". He continues by saying that the future development of skill sets needs to emphasize attributes and skills humans are better positioned than machines, like communication and interpersonal skills, problem solving, learning by doing and experimentation.

The World Economic Forum performed an extensive survey aiming at understanding the current and future impact on e.g. skill set (across all industries) in the 4th IR [32]. The information reported in Table 1 refers to skill demands in 2020 and what is thought to be a growing skill demand (2015-2020). Thirty-six percent of the respondents rated complex problem solving as a core competency (Table 1). In general, social skills are ranked higher than narrow technical skills. Content skills, cognitive abilities, process skills and critical thinking are thought to be a growing part of the core skills. Schwab [3] also argues that there are occupations that are most and least prone to automation listing what occupations likely to be "replaced" by automation. Of interest from a shipping perspective, Marine Engineers and Naval Architects rank number 8 on the list for occupations least subjected to changes through automation.

Certification and Watchkeeping for Seafarers STCW Code to be fulfilled in the digital age [35];

"The skills and competence of seafarers are ensured through the provisions of the 1978 STCW Convention and Code, as amended, which concentrate on the following functions:

- Navigation;
- Controlling the operation of the ship and care for persons on board;
- Cargo handling and stowage;
- Marine engineering;
- Electrical, electronic and control engineering;
- Maintenance and repair; and
- Radiocommunications" [35];

Table 1 Core work-related skills demand in 2020 and predicted changes for 2015-2020 (Adapted from Future of jobs [32])

Core work-related skills	Scale of skill demand in 2020	Core work-related skills	Growing skill demand
1. Complex problem solving	36 %	1. Cognitive Abilities	52 %
2. Social Skills	19 %	2. Systems Skills	42 %
3. Process Skills	18 %	3. Complex problem solving	40 %
4. Systems Skills	17 %	4. Content Skills	40 %
5. Cognitive Abilities	15 %	5. Process Skills	39 %
6. Recourse Management Skills	13 %	6. Social Skills	37 %
7. Technical Skills	12 %	7. Recourse Management Skills	36 %
8. Content Skills	10 %	8. Technical Skills	33 %
9. Physical Abilities	4 %	9. Physical Abilities	31 %

When discussing changes in training, it is necessary to take into account the time lags between input and output in the educational system. Changes in vocational training is longer and slower moving process than the developments in work life [31]. Little research has been performed asking what skillset our future seafarers need to have to operate future vessels. Given the perspective IMO and classification societies have, we will have a mix of different vessels operating under different levels of automation [25, 26, 27, 28]. It is obvious that for quite some time, during the transition towards fully automated vessels, there is a clear challenge between maintaining the traditional skill set needed to operate older vessels and trying to understand what skill sets the future operators (onboard and/or ashore) are going to need. IMO through Ashok Mahapatra, Director of Maritime Safety Division, however, refers to the need of the International Convention on Standards of Training,

The Maritime Academies are obliged to fulfil the requirements in STCW Code [34]. The situation is however changing fast and educational requirement described in STCW remain traditional to a large extent. It is hard to argue that the "core knowledge" of navigation and marine engineering in ship's operations will not be valid in the future in one form or the other. The question to ask is if it is enough, will future seafarers need additional skills and of so, what skill set needs to be added?

There seems to be an agreement within the shipping industry that human factors competencies will be needed for the crews to meet the future demands in the coming digital age. IMO recently recognised this and incorporated this into the revised version of the STCW Code, Manilla Amendments [34]. Necessary skills and competencies for leadership, teamwork and managerial skills were added and applied for both navigating and engine officers. This is partly in line with the conclusions made earlier that soft skills are identified as future core competencies in the 4th IR [32, 3]. KNect365 [37] asked 9 experts to discuss the skills seafarers is going to need in the future. Soft skills like leadership, personalized lifelong training, management but also the technical skills necessary to use the most modern equipment were mentioned which align with other industries conclusion. Comments are also made about the challenge of training for the future need we don't really know what it is going to look like.

4.1 HOW TO MANAGE THIS PROBLEM

More research is this field clearly is needed. The discussions made by different stakeholders in the industry points out that there is a gap between today's skill set and what the future seafarer is going to need. As said earlier, we are designing for a future we still don't know what will look like; we only know it is coming and that with certainty will need a different skill set, shipping no exception [3, 32]. Do try and determine necessary competencies has been researched in different areas [33]. One perspective considered in military research in Sweden is to compare the action and the competencies required for the task (Figure 2). The utilized competence builds on two pillars, represented by the individual and the task.

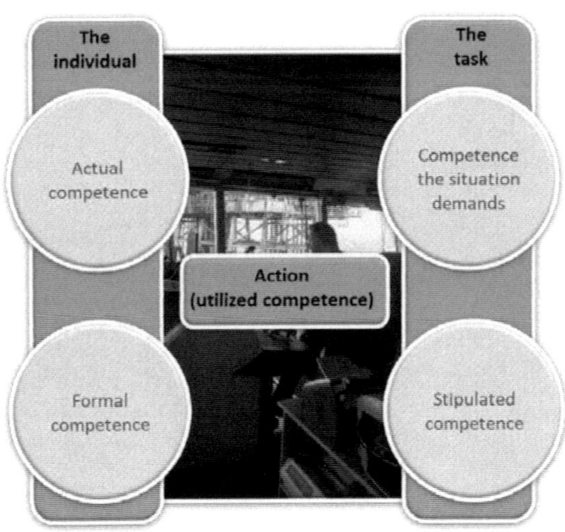

Figure 2 The 5 dimensions of competence (Adapted from [33])

Actual competence refers to the individual itself and represents the individuals gathered competence. The formal competence is competence accrued through different educational setups and can be verified by marks, certificates etc. Stipulated competence is the official requirements determining the contents of the education, in shipping represented by STCW Convention.

Rapid changes in technology and systems will likely outpace our understanding of the skills and competencies of tomorrow's maritime professional. Like the aviation and nuclear industries, the world of shipping is evolving into a much more complex and highly organized socio-technical system. Skills and competency directives must become embedded in life-long learning. There will be a need for crystal ball gazing, identifying emerging technologies before they come into practice.

5. CONCLUSIONS

A fully autonomous electric powered container vessel may be a reality in four years. Although an awareness of the need for changes and intense work effort put into adaption and regulatory support more research needs to occur. Large research projects are ongoing, and forums are established to manage the challenges the industry has ahead [38, 39, 40] Trial areas are established in Finland, Norway and China, the necessary technology is here and constantly being refined. The human element is frequently mentioned [41, 42, 43] as an important feature and within this lies the skill set. Resources and research need to be spent on trying to understand what future seafarers, "Seafarers 4.0" will look like. What skill set do they need, how are they organized and how should the be educated. Although STCW certainly implies a part of the core competencies the transformation into the digital age will require additional skills and we need to try and understand what necessary utilized competence our future seafarers will need.

6. ACKNOWLEDGEMENTS

The authors would like to thank the Swedish Maritime Mercantile Foundation and Västra Götalandsregionen for their financial support in this work.

7. REFERENCES

1. Xu, M. David J. M. & Hi Kim S., "The Fourth Industrial Revolution: Opportunities and Challenges", International Journal of Financial Research Vol. 9, No. 2; 2018

2. Prisecaru, P., "Challenges of the fourth industrial revolution, Knowledge Horizons – Economics", Vol 8, No 1, pp. 57-62, 2016

3. Schwab, K., "The fourth industrial revolution", Crown Publishing Group, ISBN: 9781524758868, 2017

4. Veiga, J. L., "Safety culture in shipping", WMU Journal of Maritime Affairs, NO.1, 17-31, 2002

5. Kobylinski, L "Marine transport and the fourth industrial revolution" Prace Naukowe Politechniki Warszawskiej. Transport, Vol. 111, Pages 269--278 (2016)

6. Bainbridge, L "Ironies of Automation" Automatica Vol 19 No. 6 pp. 775-779, 1983

7. AAWA, "Redefining shipping – A transition", https://www.rolls-royce.com/~/media/Files/R/Rolls-Royce/documents/customers/marine/ship-intel/5%20%20Business.pdf (Visited 2018-08-23)

8. Rolls Royce Ship Intelligence White Paper, "Remote and autonomous ships The next step", https://www.rolls-royce.com/~/media/Files/R/Rolls-Royce/documents/customers/marine/ship-intel/aawa-whitepaper-210616.pdf, 2016 (Visited 2018-08-23)

9. Lloyd´s Register, "Global Marine Technology Trends 2030 – Autonomous Systems" https://www.lr.org/en/insights/global-marine-trends-2030/global-marine-technology-trends-2030-autonomous-systems/, 2017 (Visited 2018-08-23)

10. Lützhöft, M "The technology is great when it works": Maritime Technology and Human Integration on the Ship's Bridge University of Linköping University, Dissertation, (2004)

11. Bloor, M., Thomas, M., & Lane, T. (2000). "Health Risks in the Global Shipping Industry: An Overview". *Health, Risk & Society, 2*(3)

12. Höivold, I. (1984). "Norwegian Research and Development in the Field of Ship Automatio"n. *Modelling, Identification and Control, 5*(3), 171-178.

13. Lundh, M., Rydstedt, L., "A static organization in a dynamic context – A qualitative study of changes in working conditions for Swedish engine officers". Applied Ergonomics Vol 55, 1-7, 2016

14. Conceição, V., Carmo, B., Dahlman, J. and Navarro, A. "Visualization in Maritime Navigation: A Critical Review". Presented at the 8th International Conference on Applied Human Factors and Ergonomics, 17-21 July, Los Angeles, USA, 2017

15. STM "Mona Lisa 2" http://stmvalidation.eu/projects/monalisa-2/ (Visited 2018-08-23)

16. EfficienSea http://www.efficiensea.org/ (Visited 2018-08-23)

17. EfficienSea II https://efficiensea2.org/ (Visited 2018-08-23)

18. MUNIN http://www.unmanned-ship.org/munin/ (Visited 2018-08-23)

19. SARUMS https://www.eda.europa.eu/docs/documents/SARUMS-Flyer-2012.pdf?sfvrsn=0 (Visited 2018-08-23)

20. MarineLink https://www.marinelink.com/news/shanghai-debuts-chinas431886 (Visited 2018-08-23)

21. Fairplay https://fairplay.ihs.com/commerce/article/4302801/stena-line-and-hitachi-europe-in-ai-partnership (Visited 2018-08-23)

22. MarineLink https://www.marinelink.com/news/samsung-heavy-industries-builds-smart-440489 (Visited 2018-08-23)

23. YARA https://www.yara.com/corporate-releases/yara-selects-norwegian-shipbuilder-vard-for-zero-emission-vessel-yara-birkeland/ (Visited 2018-08-23)

24. Lighthouse reports "Autonomous safety on vessels - an international overview and trends within the transport sector" https://www.lighthouse.nu/sites/www.lighthouse.nu/files/attachments/autonomous_safety_on_vessels_-_webb.pdf, 2016

25. IMO Maritime Safety Committee (MSC), 99th session 16-25 May 2018, at IMO London http://www.imo.org/en/MediaCentre/IMOMediaAccreditation/Pages/MSC99preview.aspx (Visited 2018-08-24)

26. The Maritime Executive https://www.maritime-executive.com/article/imo-takes-first-steps-to-address-autonomous-ships#gs.w2jgedE (Visited 2018-08-24)

27. Lloyd's Register, "LR Code for Unmanned Marine Systems" February 2017

28. Bureau Veritas, "Guidelines for Autonomous Shipping Guidance Note NI 641 DT R00 E", http://www.bureauveritas.jp/news/pdf/641-NI_2017-12.pdf, 2017

29. IACS http://www.iacs.org.uk/news/iacs-press-release-june-7-2018/ (Visited 2018-08-24)

30. Peters, M. A. "Technological unemployment: Educating for the Fourth industrial revolution" Educational Philosophy and Theory. Vol 49 No 1-6, 2017

31. Blinder, A. S. "Education for the Third Industrial Revolution" CEPS Working Paper No 163, 2008

32. World Economic Forum, "The Future of Jobs – Employment, Skills and Workforce Strategy for the Fourth Industrial Revolution" Global Challenge Insight Report, 2016

33. Försvarsmakten "Pedagogiska grunder", M7749-711000, 1996 (In Swedish)

34. IMO,"International Convention on Standards of Training, Certification and Watchkeeping for Seafarers, Manila Amendments" 2011 ed

35. IMO "Digitalization and Maritime Autonomous Surface Ships (MASS) Regulatory Scoping Exercise", https://www.dma.dk/Vaekst/autonomeskibe/Documents/Ashok%20Mahapatra%20Maritime%20Autonomous%20Surface%20Ships.pdf, 2018

36. Chawis, P. "Human Factor Competencies for the Future Mariner" Alert HE01345 Nautical Institute, 2016

37. KNect365 https://knect365.com/shipping/article/c17e1ac8-6e3a-4e33-855c-5adbb3bad9aa/9-experts-discuss-the-skills-seafarers-need-in-the-future (Visited 2018-08-16)

38. Norwegian Forum for Autonomous Ships (NFAS) http://nfas.autonomous-ship.org/index-en.html (Visited 2018-08-24)

39. AutoSea https://www.ntnu.edu/amos/autosea (Visited 2018-08-24)

40. Advanced Autonomous Waterborne Applications Initiative https://www.rolls-royce.com/~/media/Files/R/Rolls-Royce/documents/customers/marine/ship-intel/12%20-%20AAWA%20Coordinator.pdf (Visited 2018-08-24)

41. Ahvenjärvi, S. "The Human Element and Autonomous Ships" TransNAv Vol 10 No 3, 2016

42. Patriarko, D. "Autonomous ships and the human element" Maritime Journal" 11 September, 2017

43. Wahlström, M. Hakulinen, J. Karvonen, H. Lindborg, I., " Human factors challenges in unmanned ship operations – insights from other domains" 6th International Conference on Applied Human Factors and Ergonomics (AHFE 2015) and the Affiliated Conferences, AHFE 2015

44. Morris, G Risk Report:" Crewless Ships Raise Questions - Is a remote operator legally a master? New technology confounds old terms" Marine March 2018 'issue.

8. AUTHORS BIOGRAPHY

Monica Lundh is currently Head of the Maritime Human Factor Research Unit at Chalmers University of Technology, Sweden. Her main responsibility is research within work environment, human centered design and workplace design in the light of the technological changes and development that the shipping industry has and will face. She is a former Marine Engineer, served in the Swedish Merchant Navy and holds a PhD in Shipping Technology from Chalmers, Sweden.

Katie Aylward is a PhD student at Chalmers University of Technology in the Maritime Human Factors Research Unit. Her research focuses on human factors, workplace design and advanced technologies in shipping and navigation. Prior to Chalmers, she worked as a Human Factors Specialist at the American Bureau of Shipping (ABS) in Houston, Texas. She also holds a Master's of Science in Kinesiology from Memorial University of Newfoundland (MUN), Canada.

Yemao Man is a Ph.D. student of Maritime Human Factors at Chalmers University of Technology. His current research is related to interaction design and HCI in the maritime domain. His previous experience includes research regarding supervisory control system, automation, energy efficiency and service architecture. He holds a joint MSc. in Computer Science from Uppsala University in Sweden and Tongji University of China.

Scott N. MacKinnon is a Professor of Maritime Human Factors at Chalmers University of Technology. His current research is related to human-machine interactions in complex socio-technical systems. Previous to coming to Sweden, he was a Professor at Memorial University of Newfoundland where he conducted research on the impact of simulation and virtual reality upon knowledge and skill acquisition in maritime and offshore oil and gas industries. He holds a PhD in Biomedical Engineering from the University of Cape Town.

PRESENTATION

SEA U LAB: A MULTIDIMENSIONAL TOOL FOR HUMAN FACTORS EVALUATION

H. Nguma, R. Coutellier and **C. Maïs**, Naval Group Research, FR

SUMMARY

Naval Group is now equipped with a Human Factors' laboratory, named Sea U Lab and located at Naval Group Ollioules (France). The purpose of the Sea U Lab is to bring together all the methods and tools (objective and subjective measures) useful in the evaluation of existing and future solutions, in a HF frame of mind. The aim is to take into account both HMI and Workstation design and be usable for sole operator or small team (3 to 4 people) projects. Some examples of the use of the Sea U Lab are presented hereby, ranging from operational situations to applied or upstream research needs. The ultimate goal is that all new conceived systems and all upcoming innovation inside Naval Group, go through a human factors' evaluation process in order to be "HF Proven". This HF evaluation laboratory can also be used for collaboration with external partners.

NOMENCLATURE

HF	Human Factor
HMI	Human Machine Interface
HCD	Human Centered Design
EEG	Electroencephalogram
ECG	Electrocardiogram
GSR	Galvanic skin response

1. INTRODUCTION

The HF aspect of the design process of Naval Group's armed vessels and embedded systems has been taken into account for many years. This approach accommodates customer's requirement and the HF field evolution (including tools' availability).
For HF experts, it means implementing the "*scientific knowledge relative to Human and essential to the design of tools, machines and apparatus design that can be used by the greatest number with maximum comfort, safety and efficiency*" [1].

The human factors approach at Naval Group is one of participatory ergonomics with iterative validation in the different phases of intervention in the design cycles.
Prospective ergonomics, ergonomics for design and corrective ergonomics (maintaining the system in optimal operational condition) [5] are the usual application domains of HF within Naval Group. Innovative solutions must be defined and tested prior to the design of tomorrow's systems to optimize the consideration of the user.
The HF assessments carried out on systems before the creation of the Sea U Lab used to tending toward the use of questionnaires and enquiries with users.
The Sea U Lab, including its set of methods and tools, enable a multidimensional approach with the combination of objective and subjective data, depending on the assessment's goals such as cognitive workload, HMI usability.

In the remainder of this paper, we will showcase different assessments corresponding to different use cases of the SEA U Lab, putting the emphases on what the Sea U Lab bring or could bring.

2. HUMAN FACTORS EVALUATION

Human factors assessment is an integral part of HF intervention. Its aim is the assessment of the solution's utility, usability and acceptability.
The goal of HF evaluations is to avoid the rejection of a new system, to enhance the operational efficiency, to design apparatus adapted to contexts of use, needs and user's characteristics and in doing so, to enable a better quality of service.
This is done by focusing in understanding how a Human will handle his/her resources in order to cope with the situation's constraints during activity.

Depending on whether the user is involved, different types of evaluation can be distinguished and used [Figure 1]. In the case of ergonomic expert assessment, the ergonomist verifies the compliance with ergonomics principles [2] without the user's involvement. Participatory approaches can go from questionnaires, to collect users' feedback, to usability testing where the user is put in real or simulated situation to enable activity's analysis.

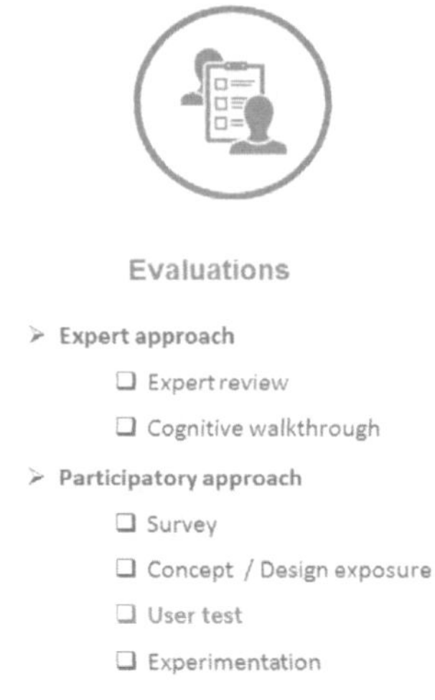

Figure 1: Different types of HF evaluations

In a user centered approach, the Sea U Lab will enable evaluation from usability testing of solutions by including objective and subjective data. These assessments can pertain to existing systems, prospective solutions and/or upstream researches.

Human behaviour's analysis is conducted by means of multidimensional assessments taking into account its cognitive, affective and biological aspects. This analysis uses physiological states, behavioural data, observation data and subjective data [Figure 2].

The evaluation methods used with the Sea U Lab are standards in the ergonomics approach, namely the observation of real world situations (activity analysis, data log, oculometry…), the experimentations (usability testing, perception testing, memory testing…), including the use of direct and indirect questioning. These methods are combined with behavioural and physiological data, in order to get objective results and to assess how a situation might affect a user's cognitive states, modifying for example the mental workload, the vigilance or the decision making.

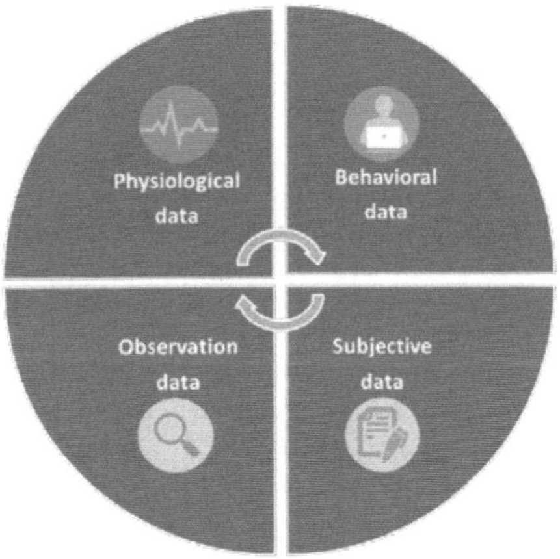

Figure 2 : Multidimensional Approach of evaluations

The choice of which method to use will depend on the evaluation's purpose, by taking into account the risks and HF challenges. Indeed, different factors have an impact on the activity including factors related to the task, to the environment, and to human variability. Before an evaluation, an analysis must be undertaken on the implications. This analysis will provide the necessary knowledge to adapt the possible HF methods and metrics. It will enable the goals definitions depending on the situation, the level of maturity and the design phase of the system. This will have an impact on the number of participants, the users' profiles needed, and the duration of the evaluation.

3. SEA U LAB TOOLS

The laboratory is made of a set of tools that enable the evaluations.
These tools can be divided into two groups. On the one hand, tools of "environment simulation" oriented toward workstation hardware and on the other hand, actual evaluation tools.

3.1 Environment simulation tools

With the help of the environment simulation tools, the user can be placed in simulated situations that are close to reality. This can be useful in different cases, as when the real situation doesn't yet exist, when the real situation is unusual and difficult to manipulate for evaluation purpose, or when the real situation is inaccessible due to other constraints, like for example operational conditions for naval systems.
The type of simulation can pertain to graphical environment, virtual environment or real-world full-scale environment [Figure 3].

- In order to study HMI, Graphical user interfaces are simulated via static or dynamic mockups, mostly through prototyping tools.
- A Computer-Aided Design software is used for simulation in virtual environment. The simulation can then be used with the help of a virtual reality headset.
- A testbed enables the simulation of a full-scale workstation. The control panel available in the Sea U Lab offers the possibility to modulate the physical arrangement of monitors (up to 4 screens). It is possible to chose the input or pointing devices, like keypad, trackball or joystick. And to change the height of the workstation. With time, it should be possible to work with a reduced staff.

The testbed can be used as a complement to a digital manikin. This will help to evaluate human posture and the visual and manual accessibility of interactive devices.

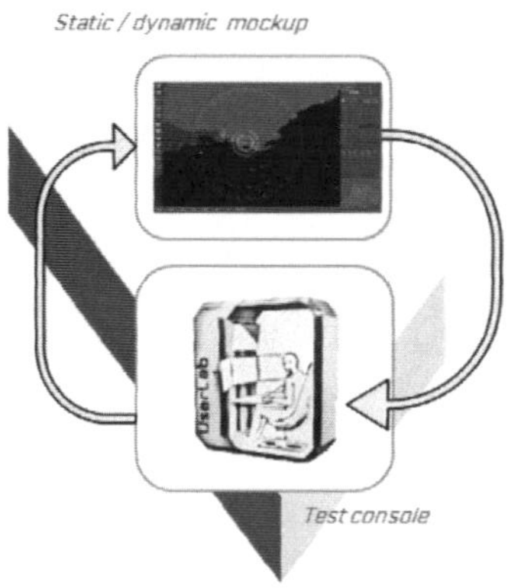

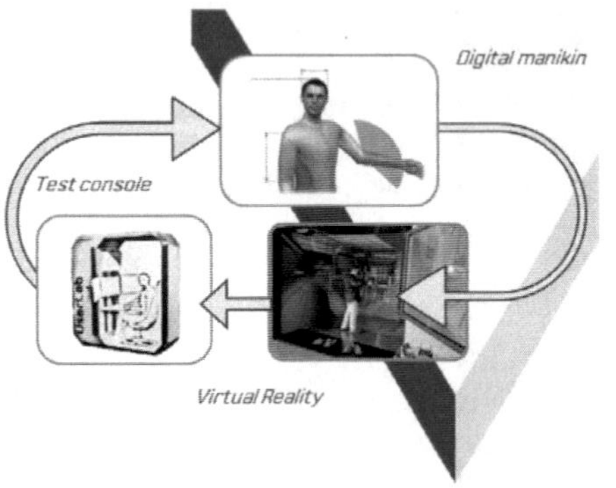

Figure 3 : Environment simulation tools

3.2 Data capture tools

Different tools of data capture, more or less invasive, are available in the laboratory [Figure 4]. They enable the capture of user's interaction with an interface, the capture of visual information gathering, the capture of user's facial expressions, or the capture of physiological data (Noldus Pack The Observer XT, FaceReader, Ulog, Media Recorder, Tobii Glasse V2, …). They are completed by a set of standardized questionnaires, and ready-to-use grids for user interview.

The keyloggers give precise and temporalized measures on the entered data, the keystrokes, and the mouse clicks on software interfaces. The recording of these objective data of Human Machine Interactions is important for any evaluation of usability and performance. For example, they enable the measurement of reaction time during action sequences realisation, but also errors or alternate paths deviating from the optimal procedure.

The tools to capture visual information enable the recording of every sessions and the real-time measurement of eye metrics such as heatmap, gaze plot, fixation time or eye movements. The recording of these data, combined with verbatims and user's behaviour data help to understand user's navigation strategies, visual information gathering and the allocation of visual attention.

The emotion capture tool offers facial expressions analysis in real time. Associated with verbatims it gives an indication on the emotions felt by the user during an interaction with a system and/or the environment.

The physiological data capture tools are of many natures. They include ECG, GSR, EEG and Eye trackers for data such as microsaccades and pupillometry. The analysis of different physiological data provides informations about the impacts of the system's use on user's cognitive states. Those states can include engagement, vigilance, workload, stress.

The use of standardized questionnaires (e.g. NASA TLX, ISA, SSQ, SUS) or the ready-to-use interview grids offer an access to users' subjective feeling.

Analysing those data separately gives informations that are open to interpretation and that may be unreliable and ambiguous. But combining objective and subjective data enable to get more reliable interpretation.

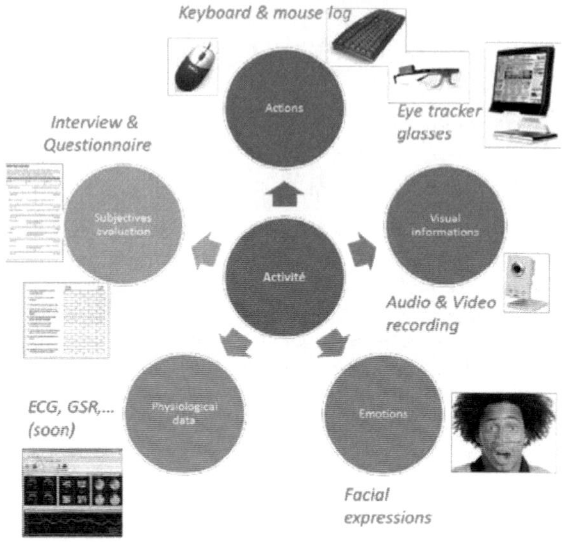

Figure 4 : Data capture tools

3.3 Data analysis tools

The data analysis is the crucial factor of any evaluation. Inside the laboratory, it is possible to use a tool for the centralization and synchronization of multisource data (The Observer XT). But, an analysis by hand is always necessary to combine all the objective and subjective data. Automatic analysis, by means of algorithms, with the help of some links between physiological states and cognitive states, enable the realization of a physiological and cognitive multidimensional analysis of the operator's state.

4. SEA U LAB USE CASES

The potential uses of the Sea U Lab are very broad, it can be used in operational situations, for upward research or applied research for innovative interfaces. Those 3 areas of the use of the Sea U Lab are illustrated below by some examples.

4.1 Example of use in operational situations.

The first example is the assessment of the prototype of a Touch-screen Tactical Table I2MAP used to present the tactical synthesis developed by the combat system. The prototype has been installed onboard of a frigate and tested by the crew.
At the end of the testing period, the ergonomist has conducted interviews focused on the utility, the usability and the attractivity of the solution [Figure 5]. These assessments, although less advanced in terms of methodology and collected metrics, provide a first insight on operating experience regarding this innovation and can

serve as the basis for the first recommendations in the definition of future products of this kind.

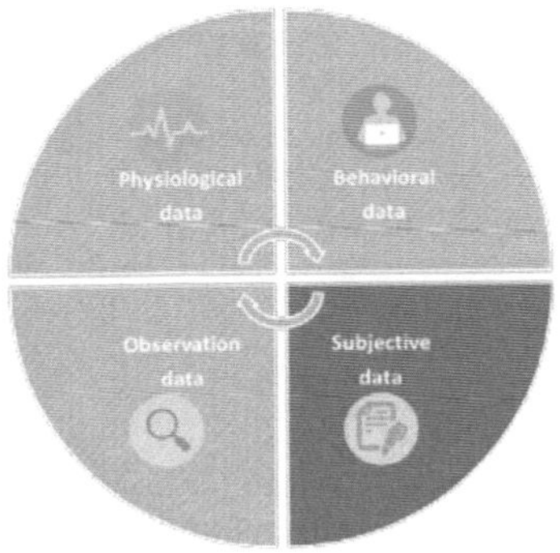

Figure 5: Methodological approach used for the evaluation of the utility, the usability and the acceptability of an innovative interface with the help of the crew members.

Given the possibility to deploy the Sea U Lab on board, the evaluation could have been completed with measures and observations better adapted to real situations [4].

4.2 Applied research: NHT, Contribution of virtual reality for a training simulator

This example is about the assessment of Naval Helohandling Training (NHT). NHT is a new educational tool that has been developed within Naval Group, it consists of a multi-player simulator for the training and development of aviation teams in the implementation of helicopters on surface ships in realistic conditions. It enables the creation of many possible training scenarios, particularly the non-compliant cases which will cause hazards. One of the goals of this simulator is to demonstrate, if necessary, the potential of virtual reality for instruction and training simulators. However, the use of virtual reality can generate side effects during the session. These side-effects may remain after the session. This virtual reality simulator sickness consist mostly in nausea, visual discomfort, and even disorientation and posture balance impairment.

The HF experimentation implemented with operators consisted of the evaluation of utility, usability and acceptability aspects and the evaluation of the virtual reality sickness in the sequencing of multiple sessions depending on the various situations [Figure 6] in term of weather and their impacts on the aviation platform of the ship: pitching, rolling and wind. For this last part, the choice had been made to combine the results from the SSQ questionnaire with those from the Romberg's Test providing data about loss of balance. The SSQ questionnaire and the posture observation have been completed and conducted before and after every session. The video recording provided additional data to the real-time observation of the posture [6].

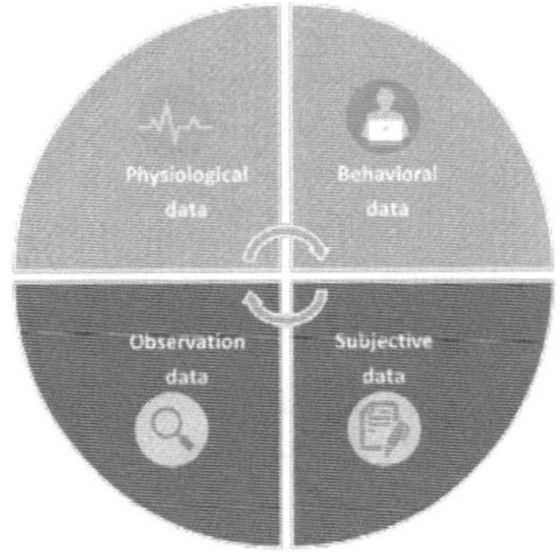

Figure 6: Methodological approach used for the evaluation of virtual reality contribution and the impact of its use on Human.

4.3 Research: evaluation of operator's behaviour

Several studies investigate operator's behaviours regarding HMI, particularly on how personality traits impact decision making, the relationship between alertness and situation awareness, the relationship between alertness and workload… In each of these evaluations the combination of observation data, behavioural data, subjective evaluations and physiological data have been relevant to get to the earliest stage of understanding how humans manage their resources [Figure 7].

One ongoing study is about the evaluation of the relationship between the synchrony effect, which is the synchronization between the chronotype and the time of day when the activity is taking place, and operator's behaviour. In order to do that, first the participants have to complete the MEQ questionnaire from Horne & Östberg [3], then they complete different detection and decision tasks. While undergoing the different scenarios, the participants are equiped with an EEG for alpha, beta and thêta waves recollection (indicators of the level of alertness), and an eye-tracker for the recollection of pupillometric data and eye fixations (indicators of the level of workload and visual strategy). The task's difficulty will be modulated to impact participants' workload.

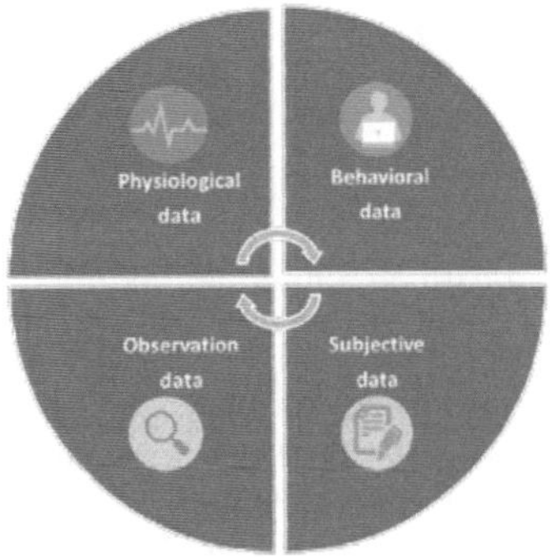

Figure 7: Methodological approach for the evaluation of operator's behaviour.

5. CONCLUSION

The usability (utility, usability, and acceptability) of embedded systems is important for Naval Group. Therefore taking into account human beings in the design process of ships and systems is a strong stake.
The human centered design approach shall allow for taking into account the users' needs depending on the task but also operator's individual differences. The multidimensional evaluation with the Sea U Lab then acquires its full meaning, be it for existing systems evaluation or for future embedded system's design. All tending towards the idea of « HF Proven » approach.
The Sea U Lab can be seen as vital for achieving this objective by allowing the objectivation of the evaluations. In view of the technological evolution (for example virtual reality creates a necessary adaptation of measuring tools) but also to obtain more accurate informations about the user, the Sea U Lab continually evolves to adjust to new requirements.

Finally, the Sea U Lab can be used (Through collaboration with research laboratories, startups …) to try the latest innovations. This will enable an upstream HF evaluation before the possible integration of new embedded systems. It will help the validation of the usefullness, usability and acceptability of innovations.

6. REFERENCES

1. Société des Ergonomes de Langue Française - SELF, 1988

2. Bastien, J. C., & Scapin, D. L. (1993). *Ergonomic criteria for the evaluation of human-computer interfaces* (Doctoral dissertation, Inria).

3. Horne, J. A., & Östberg, O. (1976). A self-assessment questionnaire to determine morningness-eveningness in human circadian rhythms. *International journal of chronobiology.*

4. Maïs, C. & Lamarca, L. (2016). Human Factors Approach in DCNS Naval Ships Process, HCI AERO 2016, 14-16 Septembre 2016.

5. Maïs C. (2016) Exploitation centrée Utilisateurs : Démarche R&D DCNS. ERGOIA'2016, 5-7 Juillet 2016.

6. Maïs, C., Lemonnier L., Bouju, Y. & Bouchet, A. (2016) Human Centered Design for an innovative training system using virtual reality: Naval Helohandling Trainer (NHT). RINA Human Factors, London. 28-29 Septembre 2016.

7. AUTHORS BIOGRAPHY

Rozenn COUTELLIER is a Human Factors specialist at Naval Group Research. She is in charge of the installation and operationalization of the Sea U Lab. Her previous experience includes both HMI ergonomics and organizational aspects for user impacts: analysis, design and evaluation in a User Centered Design approach.

Chantal MAÏS is a human-factor expert. She is in charge of research in this area at Naval Group Research. She is responsible for innovative exploitation with a Human-Centered Design. Her previous experience includes Human-Factor studies in naval ship design.

Hugo NGUMA holds the current position of PhD student at NAVAL GROUP and the university of Nice-Sophia Antipolis. He is responsible for an ongoing study on alertness and operator's behaviour.

PRESENTATION

IMPROVING MARITIME USABILITY – USER-LED INFORMATION GROUPING ON NAVIGATIONAL DISPLAYS

Viet-Dung Vu and **Margareta Lützhöft**, Western Norway University of Applied Sciences, Department of Maritime Studies, Norway

SUMMARY

As marine electronics become increasingly complex, adequate human factors consideration is necessary to develop usable systems. For navigational displays, it is required that data and control functions are logically grouped to allow the bridge team convenient and continuous access to essential information.

We present a research using card sorting to investigate the way seafarers perceive and group information. The purpose is to develop a user-led grouping of navigation data to make them easily accessible for operators.

Each participant was provided with a set of 49 cards, representing essential information elements from the displays of integrated navigation systems. The participants were asked to arrange the cards into groups based on personal judgement. Complete results from 63 participants were analysed using cluster analysis based on the level of association, allowing us to identify a grouping pattern of twelve information groups. The groups indicate a task-orientation but also artefact-related (e.g. own ship, target).

NOMENCLATURE

ECDIS – Electronic Chart Display and Information System
INS – Integrated Navigation System
IBS – Integrated Bridge System
AIS – Automatic Identification System
CPA – Closest Point of Approach
TCPA – Time to CPA
EBL – Electronic Bearing Line
VRM – Variable Range Marker
COG/CTW – Course over ground/ Course through water
SOG/STW – Speed over ground/ Speed through water

1. INTRODUCTION

The development of marine technology in recent years has resulted in an increased level of automation and the use of computing technology in navigation. On the one hand, the navigators are now provided with advanced features like the display of real-time ship position on electronic charts, safety contours, or the integration of radar and chart data to improve situation awareness. On the other hand, however, these changes also increase system complexity and create new problems.

Central issues with marine electronics include lack of a common design language among manufacturers and poor usability (Rowley et al., 2006). Poor usability often results from lack of user consideration and, already in the early 2000s, mariners have reported difficulties encountered when working with new technology [1]. Additionally, the variability in design between brands makes it difficult for seafarers to familiarise with bridge equipment when joining a new ship [2]. In combination, these issues affect user performance and increase the probability of erroneous actions, as observed in both academic studies and industry reports [3-6].

Addressing these concerns, the International Maritime Organisation (IMO) has initiated a project to standardise several display features of navigation systems using inputs from seafarers, specifically focusing on Radar, ECDIS, INS, and IBS [7]. This initiative is known officially as the development of Guidelines on standardised modes (S-mode) of operation for navigation equipment. S-mode is expected to reduce diversity in design and improve ease of use for navigation systems [7].

Among the features being standardised by S-mode is the grouping of information on navigational displays.

1.1. LOGICAL INFORMATION GROUPING

SOLAS regulation V/15 requires marine navigation systems to be designed to allow the bridge team convenient and continuous access to essential information [8]. In particular, performance standards for the presentation of navigation-related information on shipboard navigational displays [9] require data and control functions to be logically grouped, and conformity to this criteria is confirmed through analytical evaluation [10].

Examples of top-level grouping for data and control functions for radar and charting applications are provided in Annex E of standards IEC 62288 [10]. Still, those are only suggestions. Additionally, no information is provided on how such suggested grouping patterns were developed.

A frequent task for users of electronic systems is locating and accessing items on the interface, such as the screen display or menu structure. A successful interface that reflects how users naturally view content can provide maximum information with minimal cognitive effort [11]. Designing such interfaces requires sufficient knowledge of the mental models employed by people when interacting with the systems.

This study contributes to S-mode by identifying how seafarers perceive and categorise information available on navigational displays when engaging in navigation duties.

2. METHOD

To achieve the stated objectives, we employed a method commonly used by web designers and software developers to construct the structure of information on a website or a software application. The method is known as card sorting.

Originally developed by psychologist Stephenson [12] to investigate subjectivity and personal perception, card sorting has been used increasingly in usability research to study how users categorise and organise their knowledge [11, 13, 14].

As the name suggests, the method involves exercises where participants are given sets of cards representing concepts or objects and asked to arrange the cards into groups based on the level of association. The results are analysed to determine grouping patterns and help understand how people organise information, especially their perception of groups of content. A sample result of one of our pilot tests is presented in Figure 1.

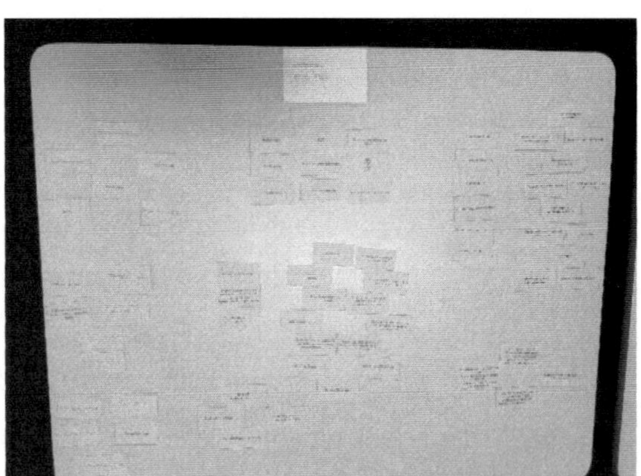

Figure 1. An example of card sorting results

2.1. PROCEDURE

Following a user-centred approach, we collected data from a broad sample of the global seafarer population. Since it is difficult to reach out to seafarers physically, we decided to conduct our card sorting sessions online using a web-based platform available at www.usabilitest.com. The only difference between sorting cards online and physically on-scene is the 'venue'.

Each participant was presented with a set of 49 cards representing essential functions and information needed for safe navigation and asked to sort the cards into groups. There was no particular requirement on grouping criteria and participants were free to choose how they would group the cards together.

2.1 (a) Selecting the cards

An important factor in a card sorting study is selecting an appropriate set of cards. The number of cards selected must be large enough to create meaningful groups and at the same time not be so high that participants get tired. In practice, it is recommended to use between 30 and 100 cards [15].

In this particular study, the number of functions and data available on standard navigation systems fits well within the 30 to 100 range. However, during our pilot tests using a set of 56 cards, we noticed that the sorting task was mentally demanding. Since the sorting would be conducted online and without facilitation, we expected that a large number of participants would drop out if the number of cards was too high. For this reason, we have taken measures to reduce the number of cards.

To select the set of cards, we examine performance standards for marine radars [16], ECDIS [17], INS [18], and for the presentation of navigation information on the displays [9] to identify features that need to be presented on the navigational displays at all times. In the end, we decided on a set of 49 cards to be used in this study. The content of these cards is listed in Table 1.

No.	Card content
1	Display mode indication for Multifunction Displays (ECDIS mode, Radar mode, Conning mode, etc.)
2	Source of Target data
3	Target Identification
4	Target Navigational Status
5	Target Range
6	Target Bearing
7	Target Heading
8	Target Course
9	Target Speed
10	Target position
11	Target Rate of Turn
12	Target CPA
13	Target TCPA
14	AIS processing status (On/Off)
15	Status of filter for AIS targets together with filtering criteria (e.g. Range, CPA/TCPA, etc.)
16	Alarm status and associated information

17	Display range scale/ Chart scale
18	Range Rings scale
19	Orientation mode (North-up, Head-up, etc.)
20	Motion mode
21	True Motion reset
22	Centred/ Off-centred display
23	Stabilisation mode and source
24	The source of own ship' speed
25	Presentation settings for target vectors (e.g., True/Relative vector, vector time, and vector stabilisation)
26	Target Trail Time and Mode indication
27	Status (automatic/manual) and level for gain and all anti-clutter control functions
28	VRM readout
29	EBL readout
30	Bearing Reference Indication (True or Relative Bearing)
31	Screen brilliance level
32	System operational status (Standby/Run)
33	Cursor Readout (Range and Bearing, geographic coordinates)
34	Chart Display mode (Display Base/ Standard Display/ All other information)
35	Chart Layer configurations (add or remove chart layers)
36	Safety contour settings
37	Safety depth settings
38	Way-point management (view/ add/ delete/ modify)
39	Route plan management (select active route/ alternative routes, store and load/ import and export routes, etc.)
40	Look-ahead (also called Own-ship check or Safety Frame – depends on ECDIS manufacturer) settings
41	Time-labels (displayed along the ship's track) settings
42	Lines of position (LOP) control
43	Route validation/Route check (Check the route plan for hazards, areas with restricted manoeuvrability, and meteorological information, etc.)
44	Own-ship geographical coordinates
45	Own-ship Heading (HDG)
46	Own-ship COG/CTW
47	Own-ship SOG/STW
48	Own-ship Rate of Turn (ROT)
49	Under-keel Clearance (UKC)

Table 1. The content of the 49 cards used in the study

3. RESULTS

3.1. PARTICIPANTS

In total, 191 people participated in the study. However, only 91 of them completed the sorting. Additionally, following the recommended practice [19], after reviewing the number of groups created by each participant, we excluded results where the cards were grouped into five groups or less to prevent the inclusion of low-quality data. This led to the exclusion of a further 28 participants. In the end, results from 63 participants remained for analysis.

3.2. METHOD OF ANALYSIS

Card sorting data can be analysed using statistical classification techniques. The most common methods are the Actual Merge Method (AMM) and the Best Merge Method (BMM), the latter of which is considered an industry standard [20]. Both methods work by merging items into clusters based on the level of association. However, AMM and BMM differ in the way items are merged.

BMM operates from the pairwise association between items. It calculates the frequency with which two items are placed together in the same group and merges the pair with the highest frequency together. This pair will then be treated as a single object to recalculate the frequency of grouping with other items, and the same merging procedure is applied for the rest of the cards. By breaking a group into pairs, BMM can accommodate for a smaller data set but has the disadvantage of loss of information by breaking up the original grouping patterns [15].

AMM, on the other hand, counts the frequency with which a particular sorting pattern has been selected by participants. When two grouping patterns have overlapping items, the pattern selected by a higher number of participants will be retained while the other will be eliminated [20]. For instance, if three participants sorted items [a, b, c, d] together and two participants sorted items [a, c, d] together, we will have two conflicting patterns. In this case, the group [a, b, c, d] will be accepted since it is selected by the higher number of participants while the [a, c, d] group will be discarded. Since AMM does not break groups into pairs, this method maintains the original grouping patterns. However, it requires a higher number of participants in order to produce reliable results.

Since we had complete data from 63 participants, which is more than twice the recommended number [21], and considering the disadvantage of BMM, we chose to analyse the data using the AMM.

With both BMM and AMM, it is also necessary to define a threshold – a minimum percentage of the participant for a selected group to be retained. For instance, with an AMM threshold of 60%, any grouping pattern selected by less than 60% of the participants will be discarded. The selected threshold is important for the resulting grouping patterns since a higher threshold increases the group number but reduces the size of each group [20]. There is no standard for selecting threshold and the value varies between studies. In this research, we observe a heuristic used in designing information architecture: for cognitive tasks, people work best with categories of five to nine items [19]. Therefore, we selected the threshold of 50% for this study to ensure the resulting groups contain between five to nine items. As a result, only groups that are agreed by 50% of participants and over are selected.

We named the items that failed the 50% threshold the "misfit" and addressed them individually. We reviewed the pairwise association between a "misfit" item and all other items of the set to determine the most suitable grouping solution.

3.3. THE FINAL GROUPS

Once the groups were set, we reviewed all labels used by participants to select a common label for each group. The final results contain twelve groups, numbered from G01 to G12.

3.3 (a) G01 – Navigation safety settings

This group contains four items:

- Look-ahead (alternatively Own-ship check or Safety Frame – depending on ECDIS manufacturer) settings
- Safety contour settings
- Safety depth settings
- UKC

The grouping of the first three items was agreed by 55% of the participants. UKC did not satisfy the 50% threshold but was added to this group because it was sorted together with both Safety contour settings and Safety depth settings by 51% of the participants, and with Look-ahead settings by 48%.

3.3 (b) G02 – Screen settings

This group contains two items, agreed by 56% of the participants:

- Display mode indication for Multifunction Displays (ECDIS mode, Radar mode, Conning mode, etc.)
- Screen brilliance level

3.3 (c) G03 – Own ship data

This group contains six items, agreed by 60% of the participants:

- Own-ship geographical coordinates
- Own-ship HDG
- Own-ship COG/CTW
- Own-ship SOG/STW
- The source of own ship's speed
- Own-ship ROT

3.3 (d) G04 – Position plotting

This group contains five items

- Bearing Reference Indication (True or Relative Bearing)
- EBL readout
- VRM readout
- Cursor Readout
- Line of position (LOP) control

The first three items were agreed by 61% of the participants. Cursor Readout did not satisfy the 50% threshold but was added to this group as it was grouped together with EBL readout by 59% of the participants, with VRM readout by 61%, and with Bearing Reference Indication by 46%. Similarly, LOP control also failed the 50% threshold but was included because it was grouped together with both EBL readout and VRM readout by 54% of the participants, and with Cursor Readout by 48%.

3.3 (e) G05 – Radar settings

This group contains two items, agreed by 63% of the participants:

- Status (automatic/ manual) and level of Gain and Anti-clutter Control functions
- System operational status (Standby/ Run)

3.3 (f) G06 – AIS settings

This group contains two items, agreed by 66% of the participants:

- AIS processing status (On/ Off)
- Status of filter for AIS targets together with filtering criteria (e.g. Range, CPA/ TCPA, etc.)

3.3 (g) G07 – Chart display settings

This group contains two items, agreed by 73% of the participants:

- Chard Display mode (Display Base/ Standard Display/ All other information)
- Chart Layer configuration (add or remove chart layers)

3.3 (h) G08 – Target data

Originally, this group contained 14 items and was agreed by 60% of the participants. However, since it is not recommended to have such a large group [22], we divided this group into smaller subgroups by raising the threshold. As a result, we have divided this group into three subgroups G08-1, G08-2, and G08-3 as presented in Table 2.

G08-1 – Supplement collision avoidance data	
Source of Target data (TT/ AIS)	Agreed by 76% of the participants.
Target Rate of Turn	
Target Identification	
Target Navigational Status	
Target Heading	
Target position	
G08-2 – Main collision avoidance data	
Target Range	Agreed by 94% of the participants.
Target Bearing	
Target Course	
Target Speed	
Target CPA	
Target TCPA	
G08-3 – Target presentation settings	
Target Trail Time and Mode indication - although this item is sorted into subgroup G08-1 by 76% of the participants, we kept this item independently since Target presentation settings are applied universally to all targets rather than any target individually, while items in both subgroups G08-1 and G08-2 are associated with individual targets.	
Presentation settings for target vectors (e.g., True/Relative vector, vector time, and vector stabilisation) – This item did not satisfy the	

50% threshold, but has moderate pair-wise association with each item of G08 (ranging from 48% to 59%). We included this item to subgroup G08-3 due to the functional association with "Target Trail Time and Mode indication".

Table 2. Target data

3.3 (i) G09 - Route planning and management

This group contains three items, agreed by 84% of the participants:

- Way-point management
- Route plan management
- Route validation/ Route check (Check the route against hazards, areas with restricted manoeuvrability, and meteorological information, etc.)

3.3 (j) G10 – Presentation settings

This group contains seven items:

- Orientation mode (North-up, Head-up, Course-up)
- Motion mode (TM/RM)
- True Motion reset
- Stabilisation mode and source
- Centred/ Off-centred display
- Display range scale/ Chart scale
- Range Rings scale

This group failed the 50% threshold and was formed after considering individual items.

The first four items in this group were grouped together by 47% of the participants. Centred/Off-centred display was included due to high association (54-70%) with each of the first four items. Display range scale/ Chart scale was included due to high association with Orientation mode (58%) and Centred/Off-centred display (59%).

Range Ring scale was included in this group since it is dependent on the selected Range scale.

3.3 (k) G11 – Alerts

Item "Alarm status and associated information" was not grouped together with any other item by more than 29% of the participants. This result suggests that this function does not belong in any other category and, therefore, could be classified into an independent group.

3.3 (l) G12 – Additional display features

ECDIS performance standards require the possibility to display time-labels and other symbols used for navigation purposes. The time-labels are specifically mentioned, and should be displayed along the ship's track on demand. The other symbols include for example points, markers and lines [17]. While preparing the cards for this study, we only included the time-labels to reduce the number of cards.

The results show that item "Time-labels settings" are not grouped together with any other features by more than 22% of the participants, which suggests that this item does not belong to any of the previous 11 groups, and therefore is classified as an independent group.

When reviewing the titles that participants used to label groups containing this item, we noticed very diverse labelling themes, including the following:

- Display settings
- Own ship data
- Route monitoring
- Chart display/ Chart settings
- ECDIS settings
- Sensor monitoring
- Tools
- History
- Vector
- Ship around zone

These diverse labelling themes suggested users have diverse perceptions of this time-labels feature, and the feature can potentially be grouped together with others. We believe a more comprehensive study using more cards will help us to specify this new group. As a result, we placed this time-label item in a new group G12, labelled "Additional display features" and intended to expand this group in subsequent studies.

3.4. DESIGNING NAVIGATIONAL DISPLAYS

The resulting groups can be used to develop a framework for organising information on navigational displays. To demonstrate this possibility, we constructed example displays for collision avoidance and route monitoring tasks. These example displays are presented in Figure 2 and Figure 3.

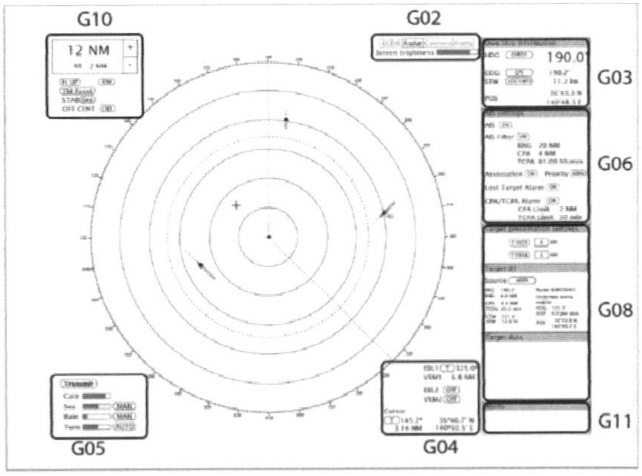

Figure 2. Example using the suggested grouping pattern for a collision avoidance display

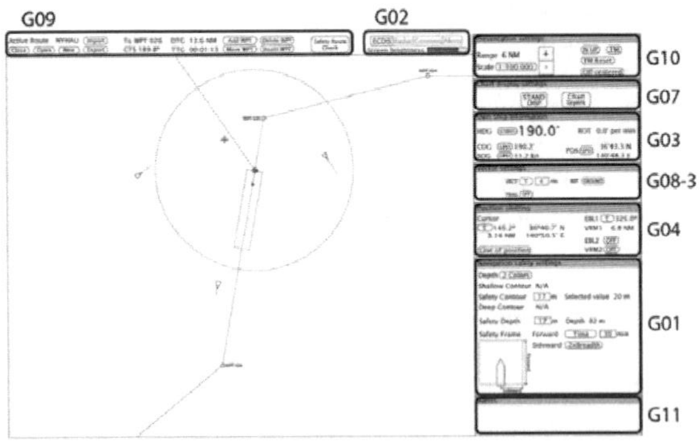

Figure 3. Example using the suggested grouping pattern for a route monitoring display

It is important to note that the displays in figures 2 and 3 only demonstrate the grouping of information, not the location of each group. Moreover, these displays are only examples and have not undergone usability tests.

4. DISCUSSION

There are limitations to this study. One is the items selected for the sorting exercises. Since the ECDIS performance standards do not provide any specific requirements regarding the presentation of mandatory ECDIS functions on the displays, many ECDIS features included in this study were selected by the research team after considering ECDIS models on the market. As a result, there may be important features absent from this study. Additionally, some non-mandatory features were excluded to maintain a lower number of cards, such as Trial Manoeuvre and Overlay features.

Nevertheless, this research has determined a common grouping pattern for essential navigation functions and data. Subsequent studies should be conducted to expand these results by including additional features.

Another limitation is the methodological issue associated with online card sorting. Since the study is conducted online without interaction between researchers and participants, we cannot identify the reasoning behind each grouping pattern. Consequently, while we can construct a grouping pattern using statistical classification, we cannot explain why mariners grouped information the way they did. For this reason, the prototype displays and the proposed groupings must also be tested with user involvement with a qualitative approach to ensure usability.

5. CONCLUSIONS AND FUTURE WORK

As marine electronics become increasingly complex, it is important that systems are designed with adequate user consideration and involvement. The aim of this paper is to understand how mariners perceive information on navigational displays, particularly their perception of grouping of content. This study has established a common grouping pattern for essential navigational functions and data. Despite some limitations, we believe the information groups presented in this study can provide a suitable framework to develop usable navigational displays.

6. REFERENCES

1. Lutzhoft, M., *The technology is great when it works: Maritime technology and human integration on the ship's bridge*. 2004, University of Linkoping. p. 49-50.
2. The Nautical Institute, *S-mode for onboard navigation displays - an NI user-led initiative.* Seaways, 2008: p. 25-26.
3. Barsan, E. and C. Muntean. *Combined Complex Maritime Simulation Scenarios for Reducing Maritime Accidents Caused by Human Error*. in *3rd International Conference on Maritime and Naval Science and Engineering*. 2010. Constantza, Romania: WSEAS Press.
4. MAIB, *Report on the investigation of the grounding of CFL Performer Haisborough Sand North Sea 12 May 2008*. 2008, Marine Accident Investigation Branch: Southampton, UK.
5. MAIB, *Report on the investigation of the grounding of Ovit in the Dover Strait on 18 September 2013*. 2014, Marine Accident Investigation Branch: Southampton, UK.
6. Rowley, I., et al., *Development of guidance for the mitigation of human error in automated shipborne maritime systems*. Southampton, UK: Maritime and Coastguard Agency, 2006.
7. IMO, *NCSR 3/28/1 Development of guidance on the Standardized (or S) Mode of operation of navigation equipment*. 2015, International Maritime Organisation.
8. IMO, *International Convention for the Safety of Life at Sea (SOLAS)*. 1974: International Maritime Organisation.
9. IMO, *MSC.191(79) Performance standards for the presentation of Navigation-related information on shipborne navigational displays*. 2004.
10. IEC, *IEC 62288 Ed.2: Maritime navigation and radiocommunication equipment and systems - Presentation of navigation-related information on shipborne navigational displays - General requirements, methods of testing and required test results*. 2014, International Electrotechnical Commission.
11. Gatsou, C., A. Politis, and D. Zevgolis, *Novice user involvement in information architecture for a mobile tablet application through card sorting*. 2012 Federated Conference on Computer Science & Information Systems (FedCSIS), 2012: p. 711.
12. Stephenson, W., *The study of behavior; Q-technique and its methodology*. 1953.
13. Doubleday, A., *Use of Card Sorting for Online Course Site Organization Within an Integrated*

Science Curriculum. Journal of Usability Studies, 2013. **8**(2): p. 41-54.
14. Dominguez, G., S.J. Hammill, and A.I. Brillat, *Toward a Usable Academic Library Web Site: A Case Study of Tried and Tested Usability Practices.* Journal of Web Librarianship, 2015. **9**(2/3): p. 99-120.
15. Spencer, D., *Card Sorting: Designing Usable Categories.* 2009, New York, USA: Louis Rosenfeld.
16. IMO, *MSC.192(79) - Adoption of The Revised Performance Standards for Radar Equipment.* 2004, International Maritime Organisation.
17. IMO, *MSC.232(82) Adoption of the Revised Performance Standard for Electronic Chart Display and Information Systems (ECDIS).* 2006: International Maritime Organization.
18. IMO, *MSC.252(83) Adoption of the Revised Performance Standards for Integrated Navigation System (INS).* 2007, International Maritime Organisation.
19. Righi, C., et al., *Card sort analysis best practices.* Journal of Usability Studies, 2013. **8**(3): p. 69-89.
20. Nawaz, A. *A comparison of card-sorting analysis methods.* in *10th Asia Pacific Conference on Computer Human Interaction (Apchi 2012). Matsue-city, Shimane, Japan.* 2012.
21. Tullis, T. and L. Wood. *How many users are enough for a card-sorting study.* in *Proceedings UPA.* 2004.
22. Miller, G.A., *The magical number seven, plus or minus two: Some limits on our capacity for processing information.* Psychological review, 1956. **63**(2): p. 81.

7. ACKNOWLEDGEMENTS

The research team wish to thank the IMO S-mode working group for facilitating the development of S-mode and, especially, the Australian Maritime Safety Authority for coordinating the group. We also wish to thank all our participants and the students at the Australian Maritime College who took part in the pilot study.

A special thanks to Optimal Workshop for providing the software for data analysis.

8. AUTHOR BIOGRAPHY

Viet Dung Vu holds the current position of PhD Research Fellow at Western Norway University of Applied Sciences. His research focuses on human-computer interaction on the ship's bridge and directly contributes to the development of the S-mode guidelines for e-Navigation.

Margareta Lützhöft is a master mariner, trained at Kalmar Maritime Academy in Sweden, with a PhD in Maritime Human Factors. Presently she holds a position as Professor of Maritime Human Factors at the Maritime Safety (MarSafe) research group at the Western Norway University of Applied Sciences. Her research interests include human-centred design and the effects of new technology.

PRESENTATION

SOLAS V/15 – BRIDGING DESIGN AND WORK

L Sørensen, BW Gas AS, BW Fleet Management and Department of Maritime Studies, Western Norway University of Applied Sciences
M Lützhöft, Department of Maritime Studies, Western Norway University of Applied Sciences
J Earthy, Marine and Offshore Technical Directorate, Lloyd's Register, UK

SUMMARY

SOLAS V/15, or Regulation 15 is called "Principles relating to bridge design, design and arrangement of navigational systems and equipment and bridge procedures". This regulation requires owners, naval architects, manufacturers and administrations to ensure compliance with specified ergonomic aims. Regulation 15 also requires owners and masters to ensure that bridge procedures are adopted which take ergonomic design criteria into consideration. At present there is no minimum manning requirement for the bridge that is linked to bridge layout. It is our assertion that such a requirement should exist which allows all tasks in the procedure/s for specific modes (ocean, coastal watch, arrival, departure, and pilot on board) to be performed. The number of staff and their competence may be adequate according to STCW, company procedures and industry best practice standards, but the bridge layout and design may not permit safe resource management. We will show that bridge layout directly impacts team work. If the bridge layout is below par, teamwork will be very demanding and at times suboptimal.

1. INTRODUCTION

Sophisticated IT-based systems are being introduced on ships (often as a regulatory requirement) without taking account of the abilities, knowledge and skills of the crew and shore staff to use this equipment [1, 2]. The International Maritime Organization (IMO) has come to realise this and the 2002 revision of the Safety of Life at Sea Convention (SOLAS) calls for changes. For example, the revised Regulation 15 of SOLAS Chapter V makes it mandatory that the bridges of new ships shall be *'facilitating the tasks...in making full appraisal of the situation...in navigating...safely under all operational conditions'*, shall be *'...effective and safe bridge resource management'*, shall be *'enabling...convenient...access...to essential information...in a clear and unambiguous manner'*, and shall be *'minimising the risk of human error and detecting such error if it occurs...in time for...appropriate action'* [3].

There are documents to support implementation; SN.1/Circ.265 [3] Guidelines on the application of SOLAS regulation V/15 to INS, IBS and bridge design, which elaborates on the V/15. Also available is ISO 8468: 2007 Ships and marine technology - Ship's bridge layout and associated equipment - Requirements and guidelines [3, 4]. MSC Circ. 982 Guidelines on Ergonomic Criteria for Bridge Equipment and Layout [5] is a complementary document, but does not consider manning, operations and training, or the cross-cutting human-centred approach to the work system proposed in the ATOMOS projects described below [6].

As part of long-term Rule development for computer systems in marine applications, Lloyd's Register (LR) were involved in the European Waterborne Transport research initiative through the five consecutive ATOMOS and DISC projects/initiatives. The strategy of ATOMOS was to enhance maritime safety and efficiency by combining evaluation of user needs with technology development. The ATOMOS IV project main theme was to develop and validate a process and tool to support equipment upgrade strategy for the European Fleet. The work that went into this was significant and innovative at the time and the authors believe it timely to remind the marine industry of their existence. In the early 2000's, Regulation 15 cut across supply chain and regulatory boundaries, and was considered a response to the 1997 NTSB incident report on the grounding of the Royal Majesty which called for Human System Integration in the maritime industry [7].

The ATOMOS response spans the life cycle of decisions for new build, plan approval, build survey, ISM and so on, and it addresses activities by a range of stakeholders. Most importantly, it addresses the bridge work system rather than focusing on either equipment or manning and training. A set of templates were produced for making submission statements attesting conformance with SOLAS Regulation V/15 2002, and include a pocket card for day to day changes, a form for minor changes and another for major retrofits & new builds [6]. The documents are available on the Human Element Alert! Website[1].

It was expected that the ATOMOS templates would be incorporated into existing document systems, and that conformance to the Regulation would be declared in a submission statement – the set of templates would assist the owner in this task. The approach to demonstrating conformance is described. The roles and responsibilities in compiling a submission to an Administration are described, but as there is no consensus on the form of such examination ATOMOS presumed that assessors would be sufficiently skilled in human element issues.

[1] http://www.he-alert.org, document numbers HE00070, HE00075, HE00080, HE00085, HE00130.

The development was based on a detailed analysis of the IMO activities that led up to the Regulation, and discussions with those responsible for drafting it [6], and it was confirmed to support implementation. IMO NAV acknowledged the ATOMOS templates as the best and most complete support for implementation of V/15.

The seven aims of Regulation 15 cannot be used directly for specification or assessment during design and development. However, ergonomic criteria can be specified that would enable these aims to be achieved. The ATOMOS approach specifies ergonomic criteria and is risk-based. A completed template is an 'operability case' concluding with an ALARP statement [7].

In short, SOLAS V/15

- Requires owners, naval architects, manufacturers and administrations to ensure compliance with specified ergonomic principles.
- Requires owners and masters to ensure that bridge procedures are adopted which take ergonomic criteria into consideration.
- If the bridge layout is below par and crew numbers at a minimum, teamwork will be very demanding and at times suboptimal.

According to SN.1/Circ.265:

The system should facilitate the tasks to be performed by the bridge team and pilot in navigating the ship safely under all operational conditions. The physical arrangement of the bridge systems and the presentation of information should permit observation or monitoring by all members of the bridge team and pilot.

Furthermore, the system and the physical arrangement should avoid the potential for a single-person failure, facilitate the bridge team and pilot in maintaining a full appraisal of the situation, and promote safe and effective exchange of information amongst the members of the bridge team and with pilots. In addition, the system should comply with appropriate ergonomic standards, e.g., Guidelines on Ergonomic Criteria for Bridge Equipment and Layout, MSC/Circ.982.

Regarding support to tasks, the circular declares that the system should allow the operator to decline or override the automated ship control functions at any time, and that the design of the system should, therefore, support the use of all means and their correlation. Furthermore, the system should be designed so that its operation minimizes distraction on the bridge that may interfere with the vigilance of the bridge team and the pilot. The focus should be on handling the ship rather than on operating the system, and the system should support the bridge team and the pilot in navigating the ship safely under all operational conditions. Finally, the system and its physical arrangement should support team working, including the assignment of tasks among the bridge team and pilot.

There is a growing amount of research on the bridge, including more recently research taking into account HCD [8, 9, 10, 11, 12, 13, 14, 15, 16, 17]. Similarly, the literature cites studies in areas such as the engine room [18, 19] and VTS [20, 21]. It is, however, clear that there is some way to go yet, especially when taking a systems view or a complete ship view. It could be asserted that naval architects know very little about shipboard work and seafarers [22, 23] although there are a few positive exceptions of applied HCD in shipbuilding [24, 25].

2. THE FIELD STUDY

2.1 RATIONALE AND BACKGROUND

Despite the advancements noted in navigational aids, standardised competency standards (STCW) and compliance with industry best practice, navigational incidents (e.g. grounding, collisions or allisions) and near misses (e.g. close quarter situations) remain a common industry occurrence. A qualitative field study was performed using cognitive ethnographic techniques to explore the impact on team work of the bridge design and layout. As part of the project the layout of the bridges were studied and analysed for possible contributions to risk and/or safety.

2.2 METHOD

Three ship visits were conducted by 3 researchers on LNG/LPG vessels between February and May 2017. Two of the researchers hold Master's licences and have commanded merchant ships and supply vessels respectively. One of the researchers is a Chief Maritime Instructor for a reputed maritime training centre. The other two researchers hold PhDs in Human Factors. The on board visits lasted between 6 and 9 days. The methods were observation, contextual interviews and semi-structured interviews, all performed on board.

2.3 PROCEDURE

Observations were performed during:
- Pre-departure preparation
- Coastal voyages
- Pilotage; including to anchor, to berth and from port
- Port entry and exit
- One bridge watch observed for all officers

Each researcher was responsible for observing a subset of the navigational operation with the associated roles performing these. As an example, one researcher observed the Master's execution of his tasks and his interaction with the bridge team, including any Pilots. Another researcher was then responsible for observing the actions of the navigating officer on duty. In addition, each researcher had a principal area of responsibility; competence, design and bridge team.

During observation of the execution of tasks researchers did not ask questions and took care not to disrupt the proceedings. Only handwritten notes were taken to minimise disruption to the bridge team. Immediately after the completion of a discrete navigational operation the researchers collated separate notes and discussed key observations made. The first ship visit was used to formulate categories pertaining to: design interaction, navigational aids, bridge team management and competence. Subsequent visits used these categories to guide observation.

3. ANALYSIS

Thematic coding of the observations was developed into six broad categories:

- Navigation practices
- Performance of bridge team
- Use of checklists and procedures
- Competency of officers and ratings
- Understanding of navigational equipment
- Bridge design

This paper presents the findings on two of the six categories, performance of bridge team and bridge design, to highlight how the team's ability to perform optimally is influenced by bridge design. The analysis of the coded observations was aided by the application of a systemic accident model to the interconnected influences of latent conditions, technological glitches and failures, human performance variability and impaired or missing barriers (Figure 1).

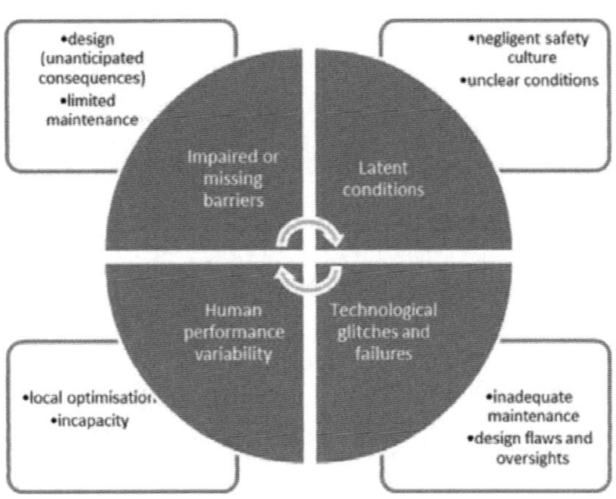

Figure 1: Systemic Accident Model (adapted from [24])

A link analysis was applied to explore the impact of the bridge design and layout on the execution of tasks by the bridge team, and a simplified visualisation is shown in figure 4.

4. RESULTS

The most important tasks during arrival and departure (including pilotage) are communication, anti-collision and navigation (anti-grounding). Communication is within bridge, within ship, between ships (including tugs and pilot boat) and ship to shore (including shore stations such as VTS and pilot station, port, linesmen, and charterers). The number of staff and their competence may be adequate, but their roles may not permit safe resource management. Design and layout of space may make task sharing difficult or impossible.

4.1. BRIDGE LAYOUT

Each vessel visited had a different bridge layout and different equipment provided. One of the vessels had a very basic bridge which illustrates the points above well. The bridge had no integrated bridge, but four separate consoles (2 radars, the starboard one not used, 1 ECDIS, 1 steering stand), and engine controls only available on the starboard side (Figure 2 & 3). During an arrival which was studied in detail this bridge was staffed with Master, Helmsman, Lookout, Cadet, Officer of the Watch and an off-watch officer.

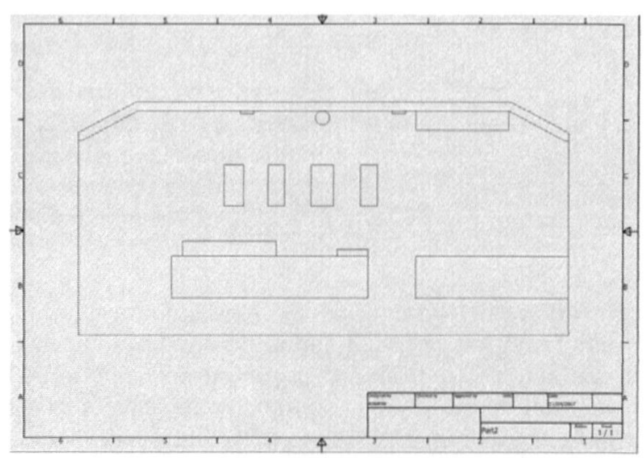

Figure 2: Bridge 1 - schematic.

Figure 3: Bridge 1.

The Link Analysis of Bridge 1 (Figure 4) shows the influence of arrangement of equipment on, for example, the team's ability to keep safe navigation using all appropriate and available means, concurrently with VHF communication, closed-loop communication in the bridge team, monitor orders to, and actions of, helmsman, giving or performing engine orders, alarm management, following procedures and all the while filling out checklists as dictated by the authorities.

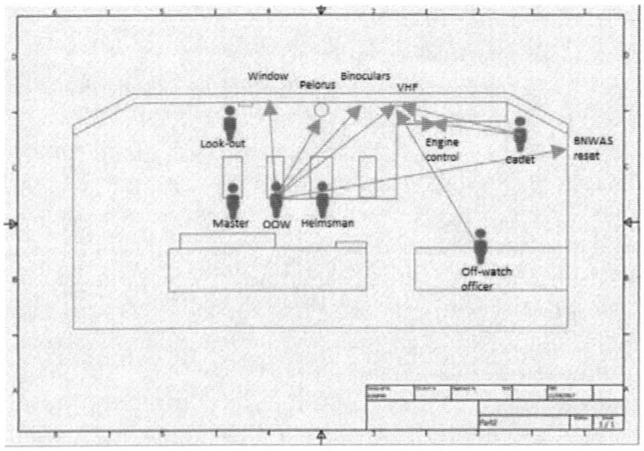

Figure 4: Simplified link analysis of Bridge 1.

The Master was mainly positioned at ECDIS checking position, issuing order follow-ups and reminders for junior officers. This means that the Master's overview of the radar was impaired.

To execute engine orders, the Officer of the Watch or the Master must cross the length of the bridge, leaving another station to do so. Thus, there was no continuous watch on radar/ECDIS.

One officer moved constantly between radar, window, pelorus, binoculars, engine, while acknowledging alarms to obtain required information. The VHFs were installed low under the front windows, and would have needed one dedicated crew member, but was now answered by anyone, meaning several crew members moved towards it (and away from their station) whenever the ship was called.

The bridge environment was very noisy with a multitude of alarms sounding at different times, paired with constant VHF chatter from nearby vessels, traffic management and port authorities (when near port). The need to acknowledge and silence alarms led to two effects: officers remained close to the equipment whenever possible so as to respond to alarms; and, increased frequency of running to equipment, either to respond to sounding alarms or to execute orders, when they were not next to them. Examples are:

- The engine control buzzed after each new order.
- Master and 3rd Officer were using portable UHFs giving noisy feedback from operations taking place on deck, cargo control room, mooring decks and elsewhere on board.
- The VHF radio was very busy and turned up loud for all to hear messages from nearby traffic and authorities.
- There were several GMDSS alarms during departure.
- The bridge had no easily visible gyro repeater, so the Master continuously had to ask helmsman for the heading. Consequently, it not only added to the noise but the bridge crew could not monitor the effective execution of the helm orders independently of the helmsman.
- A couple of times the BNWAS sounded and had to be acknowledged on the far starboard side.

Certain aspects of the bridge team performance on bridge 1 can be connected to improper team work, such as the lack of designated communication personnel or the diverse duties assigned to the officer of the watch. However, the primary observation is that the general layout of equipment and the bridge design undermined effective Bridge Resource Management. The size and layout of the bridge did not support teamwork, and the constraints on teamwork were further exacerbated by the constant follow-ups and interruptions and no closed-loop of orders – which constitutes a lack of bridge resource management in an already challenging work place. The bridge described above is contrasted by the design of Bridge 2, illustrated in Figure 5.

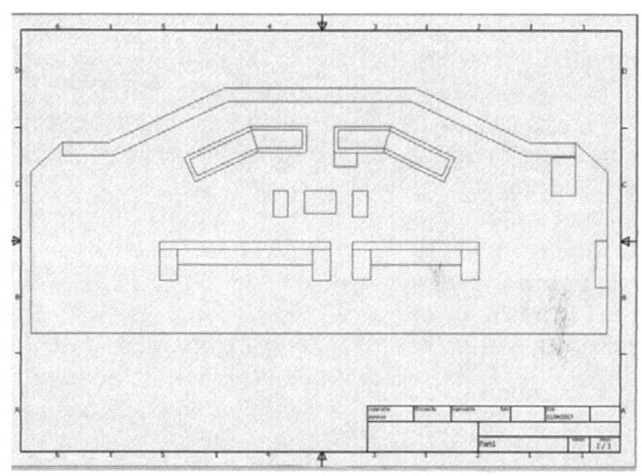

Figure 5: Bridge 2 – schematic.

Team work on the second bridge is facilitated by co-located navigation equipment such as ECDIS, radar and VHFs. The location of the engine controls is also in close proximity to the ECDIS and radars allowing the officer executing engine orders to not have to take more than a few steps away from the ECDIS or radar in order to execute the Master's orders. The Master similarly is supported in keeping an overview of the situation and the work of each member of the bridge team. The heading display is overhead in a clearly visible position, meaning the Master has no need to confirm the heading with Helmsman, thus then could act as a barrier against incorrect execution of helm orders by the Helmsman.

5. DISCUSSION

Observation of activities on Bridge 1 finds a negative impacts of equipment layout on the bridge teams ability to effectively conduct BRM, indeed the observations highlight that the layout of some equipment hindered optimal execution of tasks. The scattered allocation of equipment makes it difficult to manage bridge resources and forces several members of the bridge team to regularly move between stations, which can potentially cause disruption to watch keeping duties as well as obstruct the cooperation between members. Combined with other ergonomic issues such as the high number of alarms and the lack of a clearly visible heading indicator, the bridge design hindered the bridge team's performance. The team were forced to find ways of mitigating these effects by actions such as some team members being largely stationary by radar and others having to run around to execute orders, respond to sounding alarms and so on. Interestingly, these patterns were not observed on the bridge illustrated by Bridge 2.

The contrasting designs of Bridges 1 and 2 highlight that layout of equipment and consoles within the bridge have an observable impact on the ability of bridge teams to operate effectively. Within Bridge 2 the Master can position himself towards the back of the bridge, with access to 1 ECDIS and radar whilst seeing necessary navigational equipment unobstructed. This supports the development of Situation Awareness and enables a more effective choreography of team activities by the Master. Similarly, the members of the team can respond to task requirements as they arise; such as engine orders which can be executed in close proximity to the ECDIS, either by the Master or by any competent team member, such as the navigating officer on duty.

No difference was observed between the teams in their competence and there was no notable difference in time at sea between the different ranks. As an example, both bridge teams were headed by very experienced Masters and both had Chief Mates with similar sea time in rank. The effects discussed here are therefore seen as arising from the bridge layout and design. The analysis shows that the impact of bridge layout and design on team performance is such that adequate attention to these issues needs to be given when designing vessels and when operating them. To support these diverse requirements a set of recommendations were prepared.

6. RECOMMENDATIONS

Current regulations do not stipulate a minimum manning requirement linked to bridge layout and design. This means ship owners and ship managers must interpret the regulations in light of the bridge design and layout; noting both its constraints and its facilitation. As this paper has shown this is an area that may not have received the required level of attention it deserves. Manning requirements appear to be created as a "one size fits all" approach; e.g. in such a way that all Tankers are operated with a minimum manning level in accordance with articulated regulation, charter requirements and flag rules. This paper has shown that such a "one size fits all" approach do not adequately consider the challenges posed to BRM ability in bridges with poor layout or design. Conversely, a vessel with a well-designed bridge may be possible to sail with fewer personnel, giving opportunities for additional training positions instead (e.g. an additional Cadet or a Trainee Master).

We consider our results in the light of the ATOMOS IV SOLAS Regulation V/15 Template – Retrofit and Newbuild (available on www.he-alert.org[2]). This was chosen as the considerable changes needed would not be covered by the minor change template. The retrofit and newbuild template is the most comprehensive template and it was not used in its entirety. The risk potential was not assessed for each individual recommendation. However, it is partly evident from the previous analysis. On this note, clear statements from Flag Administrations on how conformance will be assessed are required. Without such statements, it is likely that risk-based elements will not be undertaken [6]. Naturally, full use of the template would be recommended if the task at hand was retrofit or newbuild of a ship. Figure 6 shows the activities, workflow and responsibilities for a risk-based design using the template. Based on our findings a limited set of recommendations are made.

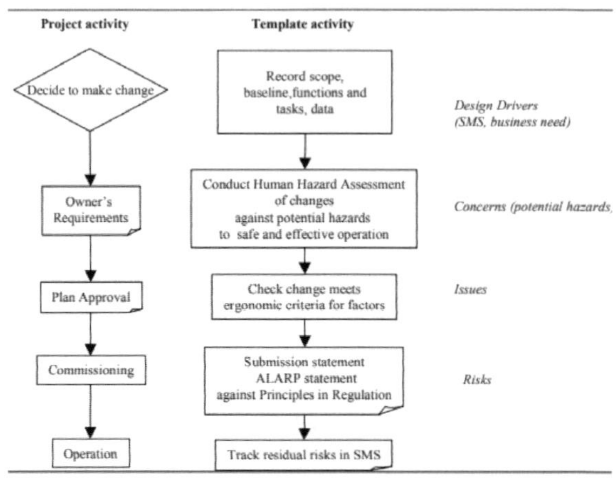

Figure 6: Flow chart for completing template in conjunction with new build or major retrofit (from ATOMOS with permission).

The steps in completing the submission statement are as follows:

1. Scope statement, when the project starts. Describe the decision being made, its rationale and scope.
2. Assess the risk potential, perhaps related to owner's requirements: Consider how the decision could affect safe and effective bridge operations.
3. Review the implementation of the decision against ergonomic criteria for the factors; training, manning and operation, equipment and SCC layout.

[2] http://www.he-alert.org/en/utilities/download.cfm/fid/E3B0A183-777B-4E0F-B273CF04D79EB0F0

4. Summarise any residual risks against the aims in the Regulation, and complete the ALARP statement.

The following recommendations are not a complete set, due to the limited scope of the study.

A. Design stages
1. Newbuilding of vessels should include (examples)
2. Risk potential (not formally assessed)
3. Review of the following circumstances
   a. Make equipment and bridge layouts adhere to a minimum standard.
   b. Alarm management.
   c. Run scenarios in mock-up environments to test and develop company specific standards for manning levels.
4. Residual risks (not formally assessed)

B. Layout vs manning
1. An assessment of vessels bridge layout and design should be included in manning level decisions.
2. Risk potential (not formally assessed)
3. Review of the following circumstances
   a. Manning levels should be set based on the number of personnel that allows all tasks in the procedure/s for specific journey segments (ocean, costal watch, arrival, departure, and pilotage) to be performed
   b. Manning levels should be taken into account when taking into account the bridge design and layout of essential navigational equipment.
   c. For vessels with a poor bridge design, an additional deck officer should be considered.
4. Residual risks (not formally assessed)

C. Workflow vs. layout
1. Shipbuilders, ship owners and ship managers should take into account exacerbating circumstances for safe and efficient workflow:
2. Risk potential (not formally assessed)
3. Review of the following circumstances
   a. overlapping communication,
   b. overlapping tasks, leading to task shedding, forgetting or delay,
   c. alarms, disturbing concentration and communication, needing acknowledgement and/or action,
   d. changes to plans requiring changes to e.g. route plan in ECDIS,
   e. task division requiring many or long locations shifts for crew.
4. Residual risks (not formally assessed)

7. CONCLUSIONS

Observations of bridge operations find the design and layout of the bridge to have a significant impact on the bridge team's performance. If designed improperly, the bridge layout can make bridge resource management difficult, cause disruption to watch keeping duties, restrict task sharing between bridge team members, and ultimately prevent the bridge team from operating effectively. Owners must take into account this potential impact of bridge design when ordering new builds. Concerning manning level, ship managers must ensure that the size of the bridge team is sufficient to perform all tasks effectively, considering the bridge design and layout of navigation systems.

Because the Regulation is in Chapter V of SOLAS, its scope is constrained to Navigation. On a ship's bridge today many diverse tasks are performed and the template is sufficiently general to address the full scope of ship control [7].

> *"The aims set out in the Regulation are difficult or perhaps impossible to assess directly"* [7] p 13.

Any changes must show an improvement in safe and effective operation, and would also need to be seen in the context of the company Safety Management System and its obligation for continuous improvement [7]. The 'operability case' approach proposed in the ATOMOS templates supports this assessment by showing the criteria in context. The Rationale document [6] contains tables listing the aims of the regulation and connecting them to product and process characteristics as a guide – as well as to training, manning, equipment and space design, respectively.

This may sound time-consuming and expensive. However, full analysis is only required when improvements in performance or safety are necessary, or when making decisions regarding novel features. Demonstration that equipment meets ergonomic criteria can be done in part during its development. During ship build the check is that items of equipment as used on the bridge do not as a whole induce errors [7].

Despite the IMO recognition of the ATOMOS work and Circ.265, we see little uptake of an ergonomic approach to bridge design. The ATOMOS work, now 15 years old, contains highly useful guidance and still-valid arguments and, despite the very large number of downloads from the HE-Alert! Website, operability cases in support of novel bridge designs are still uncommon.

However, we are hopeful because there is a growing number of good examples showing that it can be done; and that investing time, expertise and money to improve human element aspects really pays off. These include: the Tamar-class lifeboat, operated by the Royal National Lifeboat Institution (RNLI); the Service Operations Vessel 9020, designed by Damen Shipyards Group; and the Pure Car/Truck Carrier (PCTC) Harvest Leader, designed by Andreas Shipping Ltd [26].

8. ACKNOWLEDGEMENTS

We wish to acknowledge the fundamental work done by Brian Sherwood Jones in recognising and implementing the ergonomics perspective in V/15.

9. REFERENCES

1. Lützhöft M. "The technology is great when it works": Maritime Technology and Human Integration on the Ship's Bridge [Ph.D. thesis]. Linköping. Available at: http://www.diva-portal.org: Linköping University; 2004.

2. Rothblum AM, Sanquist TF, Lee JD, McCallum MC. Identifying the effects of shipboard automation on mariner qualifications and training and equipment design. ISHFOB '95: The Influence of the Man-Machine Interface on Safety of Navigation, Proceedings of the International Symposium on Human Factors On Board November 1995, Bremen, Germany. 1995.

3. IMO. Guidelines on the application of SOLAS Regulation V/15 to INS, IBS and bridge design. SN.1/Circ.265. London: International Maritime Organization; 2007.

4. ISO 8468:2007 - Ships and marine technology -- Ship's bridge layout and associated equipment -- Requirements and guidelines. 2007.

5. IMO Guidelines on ergonomic criteria for bridge equipment and layout. MSC/Circ.982. London, International Maritime Organization. 2000.

6. Sherwood Jones, B., Earthy, J. RATIONALE FOR SOLAS REGULATION V/15 TEMPLATE. ATOMOS IV REVISION WP8.5. (available on http://www.he-alert.org/). 2003.

7. Sherwood Jones, B., Earthy, J. Using an 'Operability Case' to provide assurance of safe bridge watchkeeping. The Ergonomist June 2003 (5). Newsletter of The Ergonomics Society.

8. Grech MR, Lutzhoft M. Challenges and Opportunities in User Centric Shipping: Developing a Human Centred Design Approach for Navigation Systems. OzCHI '16, November 29-December 02; Launceston, TAS, Australia. 2016.

9. Grootjen M, Neerincx MA, Passenier PO. Cognitive task load and support on a ship's bridge: Design and evaluation of a prototype user interface. INEC 2002: The Marine Engineer in the Electronic Age - Conference Proceedings; 2002.

10. Jacobson E, Lützhöft M. Developing User Needs for S-Mode. NAV08-ILA37 International Navigation Conference; London, 28-30 October. 2008.

11. Allen P. Perceptions of technology at sea amongst British seafaring officers. Ergonomics 52(10):1206-1214. 2009.

12. Petersen ES. User Centered Design Methods must also be User Centered: A Single Voice from the Field. A Study of User Centered Design in Practice. Gothenburg: Chalmers Technical University; 2010.

13. Joiner D. User Feedback in Ship Design. New Zealand: Massey University; 2007.

14. Atwell C, Pourzanjani M, Pearce D. User requirements for new navigation systems. In: Zhao J, Wilson PA, Hu Z, et al., editors. Maritime Collision and Prevention. Vol. 2. Navigation and Engineering Aspects of Collision: Chiavari Publishing p. 73-87. 1996.

15. Heasly CC, Dutra LA, Kirkpatrick M, Seamster TL, Lyons RA. A User-Centered Approach to the Design of a Naval Tactical Workstation Interface. Human Factors and Ergonomics Society Annual Meeting Proceedings 37:1030-1030. 1993.

16. Aldridge AJ, Brooks PG, Moreton MB, Smeaton GP. A user-centred evaluation for integrated bridge systems. Safety at Sea International 28-33. 1997.

17. Petersen ES. Engineering usability. Gothenburg, Sweden: Chalmers University of Technology; 2012.

18. Andersson M, Lützhöft M. Engine Control Rooms - Human Factors. RINA Human Factors in Ship Design, Safety & Operation IV; 21-22 March; London. 2007.

19. Lundh M, Lützhöft M, Rydstedt L, Dahlman J. Working conditions in the engine department – A qualitative study among engine room personnel on board Swedish merchant ships Appl Ergon. 42(2):384-90. 2011.

20. Praetorius G, Lützhöft M, Bruno K,. The context matters: Maritime safety in the Vessel Traffic Service (VTS) domain. Reliability, Risk and Safety: Back to the Future; 2010.

21. Praetorius G, Lützhöft M. Decision support for vessel traffic service (VTS): User needs for dynamic risk management in the VTS. Work 41 (SUPPL.1): 4866-4872. 2012.

22. Abeysiriwardhane A, Lützhöft M, Petersen ES, Enshaei H. Human-centred design knowledge into maritime engineering education; theoretical framework. Australasian Journal of Engineering Education 21(2). 2017.

23. Lutzhoft M, Petersen ES, Abeysiriwardhane A. The Psychology of Ship Architecture and Design. In: MacLachlan M, editor. Maritime Psychology: Research in Organizational & Health Behavior at Sea: Springer. 69-98. 2017.

24. de Vries L, Costa N, Hogström P, Mallam S. Overcoming the challenges of integration of Human-centred Design within the Naval Architecture ship design process. Proceedings 19th Triennial Congress of the IEA. 2015.

25. Bialystocki N. Human Factors Dimensions in the Design of a PCTC Vessel. International Conference on Human Factors; Royal Institution of Naval Architects, London, UK: The Royal Institution of Naval Architects; 2016.

26. Lützhöft, M., Vu, VV. Design for safety In: HA. Oltedal, M Lützhöft. (eds.). Managing Maritime Safety. Routledge, pp 106-140. 2018.

10. AUTHORS BIOGRAPHY

Dr Linda Sorensen holds the current position of Organisational Development & Human Factors Specialist at BW Gas. She is responsible for Human Factors. Previous experience includes positions as a Senior Consultant and Technical Lead at Lloyd's Register and research appointments at the University of Southampton, the University of Southeast Norway and most recently at Western Norway University of Applied Sciences.

Dr Margareta Lutzhoft is a Professor in the MarSafe group at the Western Norway University of Applied Sciences. Her research interests include Human Factors, human-centered design and the effects of new technology.

Jonathan Earthy is a Principal Human Factors Specialist for Lloyd's Register Marine and Offshore. He coordinates standards, research and assessment activities and products. His specialist area is assurance of the quality of Human Factors work, in particular for complex IT systems.

DEMONSTRATING A MARITIME DESIGN SYSTEM FOR REALISING CONSISTENT DESIGN OF MULTI-VENDOR SHIP'S BRIDGES

K Nordby, S Frydenberg and **J Fauske,** The Oslo School of Architecture and Design, Norway

SUMMARY

Maritime workplaces in digitally integrated multivendor ship's bridges often offer inconsistent user interfaces and suboptimal workflows for navigators. We propose that design consistency in such bridges can be achieved by applying methods and technologies from the web and mobile industries. We present ongoing design work directed towards defining a maritime design system enabling design consistency across multi-vendor ship's bridge systems. The system has been developed through an experimental process involving analysis of 10 user interfaces from existing ship's bridges to understand their functional structure. We further apply design strategies from the web industries, such as responsive design, style theming and standard user interface components. We present a design proposal and discuss the results in terms of consistency and the implications for industry of having a maritime design system. We argue that there are compelling advantages for the maritime industry in systematically transferring methods, processes and tools from the web industry.

1. INTRODUCTION

In a study investigating the impact of technology on safety for ship operators, Mišković, Bielić and Čulin found that 37% of the respondents had operated three or more types of integrated bridge systems (IBS) [1]. Further, they stated that 27% of the operators reported having been confused by the information provided in the systems, linking the effect to the operators encountering too many different types of equipment. Adding to these challenges is the fact that many ship's bridges do not have fully integrated IBS but are equipped with systems assembled of equipment delivered by multiple vendors. In field visits, we have observed such multivendor ship's bridge systems (MBS) consisting of up to 35 different types of equipment. These separate equipment units would be physically installed in large work consoles, leading to cluttered workplaces with suboptimal organisation of the space around the navigators [2].

Figure 1: SEAQ bridge delivered by Vard Group AS. (Photo: Vard Group AS)

In order to deliver improved workplaces, a number of MBS integrators have started to integrate equipment digitally in their workplaces. Figure 1 shows an example of one such bridge, delivered by Vard Group AS, that allow users to access equipment through touch screens. In these systems, the separate equipment units are mediated as applications with digital user interfaces that share screens and interaction devices. By *user interfaces*, we mean the access points where the user can interact with the system. These MBS are similar to IBS, but because separate companies design many of the applications, there is a great deal of variation in user interface design within each MBS.

Design consistency across applications has long been fostered in the web industry through various approaches. Here, we explore how a selection of these approaches may be applied to maritime user interfaces in order to achieve design consistency. We examine this through an experimental process in which we deconstruct the user interfaces of current ship's bridge equipment and redesign them using consistent design methods drawn from the web industry.

Our work is part of an industry-driven research project called Openbridge [3] that seeks to harmonize implementation and user experiences of user interfaces in MBS. The project consists of 20 industry, research and government partners and is led by Ocean Industries Concept Lab from The Oslo School of Architecture and Design.

1.1 DESIGN CONSISTENCY

Design consistency is one of many factors that are important in designing user interfaces. Nielsen has suggested that consistency supports the user's ability to transfer skills from one system to another and may lead to ease of learning, ease of use, higher throughput and reduced errors [4]. Design consistency can encompass many dimensions of user interface design [4]. For instance, it can relate to aspects of the graphic design of the user interface, such as spatial organisation of components, colours, symbols and typography, and to aspects of interaction design, such as structure of content, user interface patterns, terms and interaction mechanisms. Consistency can also be applied within a single application (e.g. Electronic chart display and information system (ECDIS)), across multiple applications working within a single system (e.g. an IBS) or across multiple systems each with multiple applications (e.g. across several IBS). Consistency cannot cross all aspects of an application, and consistency can in some cases result in suboptimal user interfaces [5]. Design for consistency must therefore consider users' activity.

The lack of design consistency in the maritime industry has been well known for many years, and International Maritime Organization (IMO) initiatives such as S-mode are trying to move the industry towards improved consistency across maritime user interfaces [6]. S-mode seeks to introduce a switch that would enable a standard user interface for navigation systems. Despite such efforts, current regulations and laws have not led to clear premises for consistent design across the many vendors delivering equipment for MBS. A consistent MBS requires consistency across a number of applications, and this is challenging from an organisational perspective since multiple vendors needs to agree on design guidelines specific enough to achieve design consistency. This is also problematic from a commercial perspective because user interfaces are part of both system integrators' and equipment vendors' unique value offerings.

1.2 CONSISTENCY IN WEB INDUSTRIES

In web and mobile-oriented industries (web industries), there is a long tradition of addressing design consistency through the development of design guidelines [7-9]. Developers of operating systems (such as Android, Windows and iOS) have tackled the problem by producing and distributing comprehensive design guidelines connected to software development resources that make it convenient for application developers to follow their design philosophy [7-10]. For operating system vendors, this move promotes consistent design that improves the usability of the applications built for their platforms. For system developers and designers, these guidelines contribute to improving their design delivery and ensure that the applications are consistent with the general platform on which they operate. This results in applications that are easier to learn at launch since users are already familiar with central aspects of their user interface based on previous experience with other applications.

Current generations of web-based design guidelines often take the form of a *design system*. By design system, we refer to a modular user interface methodology built on web technology that merges traditional design guidelines with development tools. A design system should be an adaptive system that supports a portfolio of applications and is in continuous development to respond to new needs [11]. The thinking behind design systems is not new [4]; however, it has gained traction in recent years as a way of reducing development costs and improving user experience (UX) for companies with large portfolios of applications. UX, which is often used as a qualitative measurement of success [12], refers to a user's dynamic, context-dependent and subjective experience of the system [13].

Although building a design system requires extensive and comprehensive work, it will ultimately save both time and money as the design system is applied to increasingly more applications and services. This is because a design system includes reusable components that simplify implementation and design processes. Because consistency and cost are important challenges in the maritime industry, we suggest that applying the thinking behind design systems might be beneficial in this domain.

2. DESIGN CONSISTENCY IN MARITIME USER INTERFACES

In reviewing current design regulation and guidelines in the maritime sector, we found no information about how to realise consistent design across different vendors [14]. Although maritime regulations and guidelines offer important support of screen based user interface design, they do not provide user interface guidelines with the specificity that is necessary to achieve design consistency. In addition, we found no support for achieving more cost-effective design of new user interfaces.

We argue that there is a gap between state-of-the-art web-based design support and design support specific to the maritime industry. To bridge the gap, we argue that it would be beneficial for the maritime industry to develop a design system based on approaches from the web industry to help achieve design consistency across maritime applications.

To do so, it is important to distinguish between an integration platform and applications. An integration platform includes the workstation and the software that manages how applications are mediated in the workplace. Applications are software programs (such as Dynamic positioning system (DP), ECDIS and wiper control systems) installed within the integration platform. A full maritime design system would cover both areas. However, this article addresses only consistency among applications. We will explore consistency across multiple complete MBS in forthcoming articles.

In order to develop a design system proposal, we have compared the user interfaces of 10 maritime applications:

- Compass system

- Echo sounder
- Deck light system
- 2 conning displays
- Electronic charts
- Alert system
- Automation system
- Application launcher
- Propulsion indication interface

The selected applications represent a wide range of digital user interface types found in ship's bridges. We have evaluated design consistency across these 10 interfaces and identified the components used to build them. Our analysis is divided into three categories: user interface components, user interface layout and user interface style. These represent common categories for organising web-centric user interface guidelines.

2.1 USER INTERFACE COMPONENTS

User interfaces are constructed using a set of components for input control, such as buttons and toggles, navigation, such as search field and sliders, and information, such as icons and notifications. In order to achieve consistency, the appearance and use of such components should be similar across all applications.

None of the interfaces we analysed used user interface components from a shared library; all opted for individual design. This resulted in much variation even in very simple interaction components, such as buttons. The variations extend to graphical icons and maritime user interface components, such as representations of thrusters. We found no functional reason for the variations in user interface component design.

2.2 USER INTERFACE LAYOUT

We refer to the overall structure of a user interface as the user interface layout. This includes the division of screen space into functional areas as well as rules for the position of important function categories, such as function and navigation menus.

We found significant variation in the user interface layouts of the 10 applications we analysed. The variations were related both to the separation of functions and to the positioning of function categories. Some of this variation can be attributed to many applications targeting different modes of operation and having different types and quantities of information and functionality. For instance, the user interface for the echo sounder represents a much narrower scope of information than that of the conning application. However, even in such cases, we do not see any reasons for the difference in layout with regard to the positioning of navigation menus and alert components, for example.

Nine of the applications were custom-made for a fixed screen resolution and could not be scaled to other screen formats, distances and resolutions without redesign. In some of the user interfaces, this was already a problem, as some of the applications were positioned farther away from the user than their interfaces had been designed for. We argue that a layout structure allowing adaptive scaling would be beneficial in many of the user interfaces.

2.3 USER INTERFACE STYLE

User interface style addresses visual formatting, such as typography, colour, spacing, iconography and the appearance of shapes and lines. Visual formatting is important in maritime applications because it is connected to readability and the design of day and night palettes. In addition, it conveys the visual identity of companies that make equipment and bridge systems.

Overall, we found no consistency across components, layout or style in the 10 applications we examined. We consider this lack of consistency problematic and contend that it most likely would lead to usability problems if the applications were installed in the same MBS.

3. TOWARDS A MARITIME DESIGN SYSTEM

We propose to develop a design system for the maritime domain able to support consistent design and efficient implementation of user interfaces for maritime applications, especially targeting MBS. Our maritime design system will include design guidelines aligned with maritime regulations, maritime-specific content, online support material and online development libraries supporting user interface implementation.

This article reports on processes aimed at the development of such a maritime design system, focusing on our collaboration with industry partners to develop an early concept for design guidelines to support application design for MBS. Our current proposal is not a full design system but a substantial step on the way towards this goal. The main purpose of our proposal is to demonstrate whether technologies and methods from web industries can be applied to maritime applications to realise consistency.

The proposal has been developed using an iterative design process where we redesigned existing user interfaces using concepts from web-based design guidelines. We especially draw on Google material design [6] due to its comprehensive connected support material and its familiarity as one of the best-known design systems. By iteratively designing multiple applications in parallel, we are able to identify common user interface elements and synchronise the design across applications. This process is repeated in a continuous loop as we try to achieve meaningful consistency in the components, layout and style of the user interface.

We have reviewed our design proposal through two industry workshops with partner companies and through expert evaluation by designers and human factor specialists working in the maritime domain.

The designs have been developed using common tools for user interface design, such as paper sketches, graphic design software and web development tools. We have developed a global style guide in parallel to the design of individual applications. In addition, we have implemented a few of the user interfaces as responsive HTML demonstrators to test scaling in practice. These have been installed in an MBS made by one of the project partners for compatibility testing. The current proposal is an early iteration of a design system and is not ready for industrial implementation. We consider it a framework for further design of a functional maritime design system.

3.1 DESIGN CONCEPT

Our design concept simplifies current application designs by limiting variations in visual style, layout and components. In addition, we strive to remove redundant information in the various user interfaces.

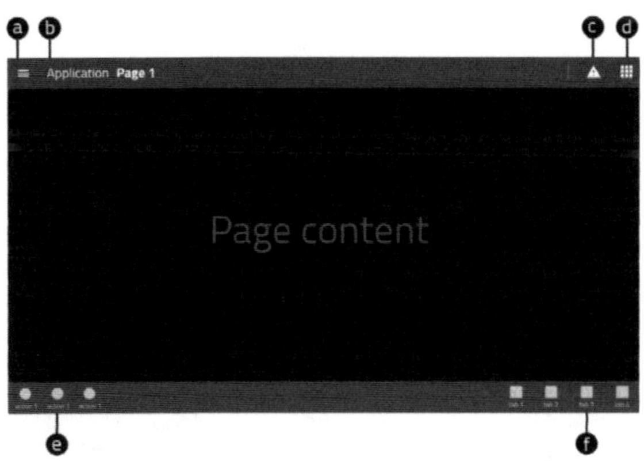

Figure 2: Main user interface window: a. hamburger menu, b. main title, c. alarm component, d. global navigation menu, e. action buttons and f. tab menu.

The proposed design system calls for application interfaces based on a simple frame with a top bar and a bottom bar (Figure 2). The bottom bar is optional if this space is needed for user interface functionality. A hamburger menu in the top left corner (Figure 2a) provides access to navigation within the application. The lower left corner has space for action buttons targeting the content area of the application (Figure 2e). The lower right corner has space for an optional tab menu providing access to application functions (Figure 3d). The top right corner has a fixed area for the alert system (Figure 2c). In addition, there is a button that offers access to a global navigation menu to swiftly switch applications and to change the palette and dimming options (Figure 2d). Central icons in the navigation menu can be placed to the right of the navigation menu button if necessary. Within the hamburger menu, the top area is reserved for navigation within the current application (Figure 3a) and settings (Figure 3b). The lower area allows access to company-specific information and the help system (Figure 3c). We have standardised the placement of important generic functions, such as dimming (Figure 2d), palette control (Figure 2d), help (Figure 3c), company messages (Figure 3c) and alerts (Figure 2c).

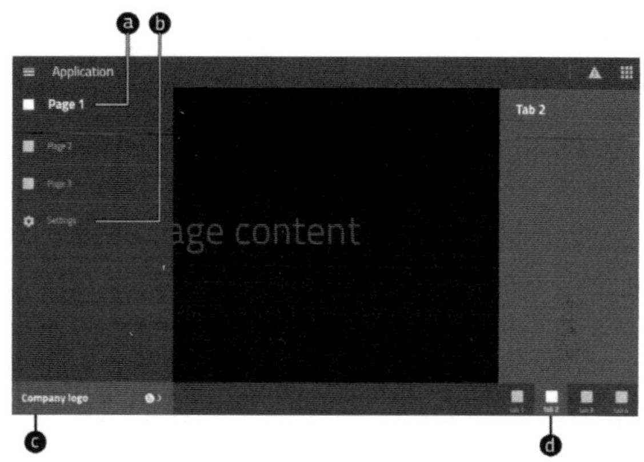

Figure 3: Main user interface window with hamburger menu revealed: a. settings button, b. navigation button and c. company-specific flyout.

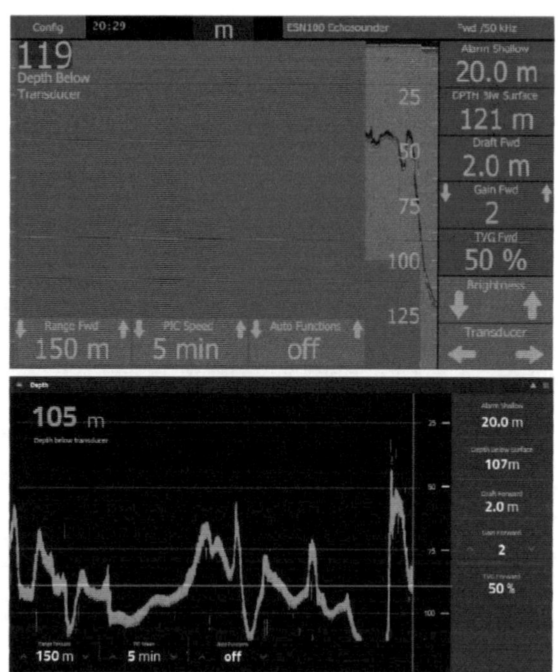

Figure 4: Comparison of old user interface (left) and new user interface (right).

Figure 4 shows an example of an actual application design and the same application converted to the new template. The new template adds new functionality, such as free scaling and optional hiding of menus. The design also allows comprehensive changes in visual style. The palette in figure 4 is drawn from some of the partner companies' application implementations. We are in the process of designing a new palette that conforms to maritime regulations.

Our design proposal makes use of several approaches found in web-based user interface design: theming, responsive design, common user interface component design and a user interface component library. What follows is a description of each of these strategies and a walkthrough of how we have implemented it in our design.

3.1 (a) User Interface Components

Although we found significant variation in the *design* of user interface components in our analysis of current systems, we also saw that the number of component *types* was quite small, dominated by buttons, toggle buttons and menus. Our design proposal makes use of the component types we found in our analysis and proposes new common components inspired primarily by material design (Figure 5).

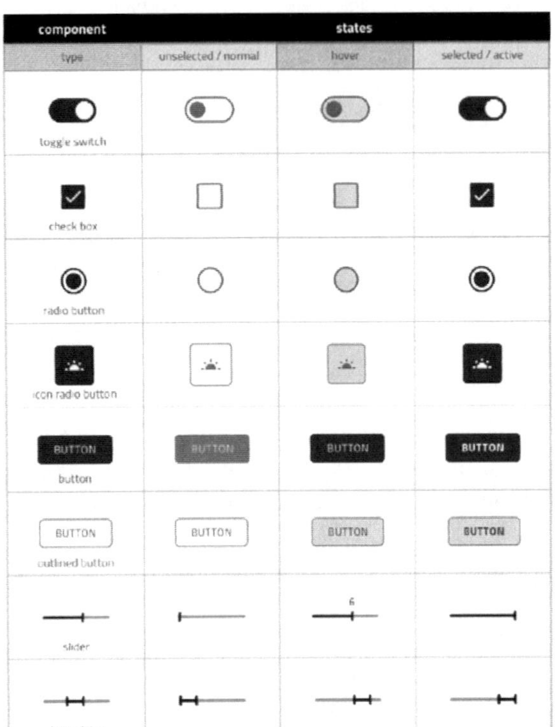

Figure 5: Generic user interface component library.

We opted to scale the buttons for touch interaction for two reasons. First, touch interaction is common in modern user interfaces for ship's bridges. Second, implications of fatigue and a moving environment may potentially affect a person's ability to identify and trigger a button, so we assume that larger hit areas are a benefit in maritime systems in general. This choice needs further evaluation in formal user tests. Future iterations of the concept may differentiate between components for touch interaction and components for screens operated through indirect interaction, such as by mouse.

In addition to user interface components borrowed from web industries, we have defined an additional category of maritime-specific user interface components. There are a number of interaction components that represent specific types of data, such as visualisations of heading and thrusters. We have identified these as *maritime user interface components* connected to specific maritime content. Also within this group are components that mediate specific generic functions in the user interface, such as the alarm component and the dimming component.

3.1 (b) Theming

It is common in web development to separate definitions of application styles from other code and to use cascading style sheets (CSS) and similar technologies to format the visual style of an application. This method makes it possible to change user interface attributes, such as spacing, fonts, colours and geometry, for any number of user interfaces by simply editing a global CSS file. Figure 6 show an example of a toggle button formatted by two palettes. Theming across an MBS would enable uniform day and night palettes and a custom style guide to accommodate individual users and would also allow adaptation of the visual style of any application to an integrator's design scheme. This approach would make it possible for an equipment manufacturer to deliver a single application that could automatically be adapted to different style guides on different MBS.

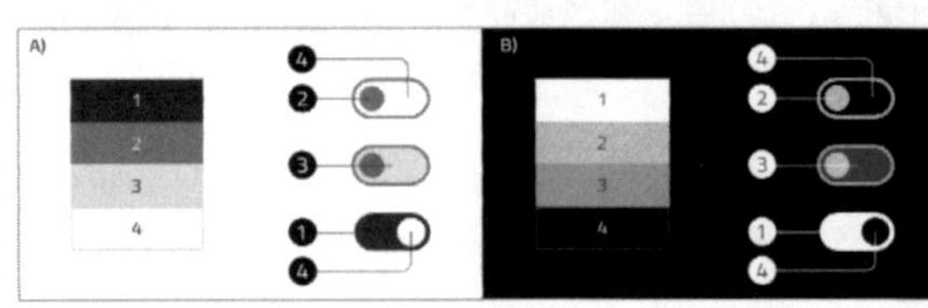

Figure 6: Colour palette connected to a toggle button user interface component. a. Light palette, b. Dark palette.

We have based our approach to theming on our partner companies' needs to adapt the look and feel of the design guidelines. To facilitate this goal, we have integrated fonts, colours, line thickness and border shape into the CSS specification. The system can be used to format a single application or all applications within the MBS. This allows us to achieve complete aesthetic conformity across workplaces and opens up new possibilities, such as allowing ship owners to implement their own style guide on an MBS.

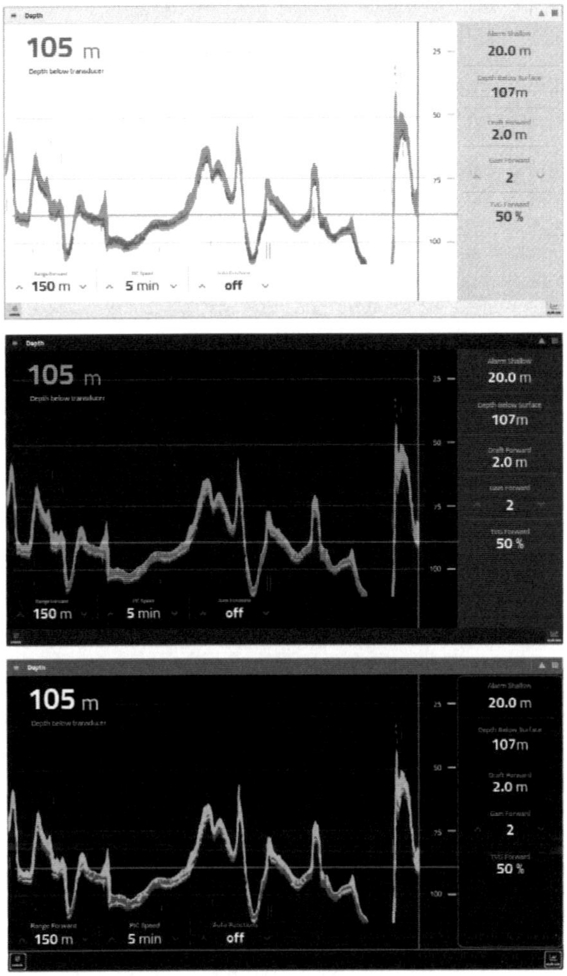

Figure 7: Comparison of three visual themes.

The lower two themes in Figure 7 are based on existing design guides used by integrators in the project group. The two designs demonstrate that we can replicate some of the visual identity of existing systems through CSS while maintaining the layout and component design of our design system.

The theming functionality of the design system is based on a taxonomy of the elements that make up the visual representation of the applications. This include lists of allowed text formatting, line types, geometry types, icon formatting and colour palettes. To ensure that all graphics can be formatted through CSS, we do not allow use of images (raster graphics) as part of user interfaces. Instead, all graphics are generated as vectors, which support colour changes using CSS while having the advantage of supporting free scaling without pixilation.

Maritime theming will differ somewhat from current web-based templates. First, the design must facilitate multiple palettes related to operational conditions, such as night, dusk and dawn palettes. Second, maritime requirements and user needs warrant the connection of some colours to specific functionality, such as alerts. Third, the system must facilitate text scaling based on viewing distance in addition to screen size.

3.1 (c) Responsive Layout

As mobile phones took over as the primary platform for consumption of web services, the web industries recognized the need for streamlined web page design that could scale across multiple screen sizes. Responsive design or responsive layout solved this problem by establishing rules to define how a web page or application would rearrange its content according to how it was accessed [15].

Many of the participants in our consortium have reported the difficulty of supporting multiple screen sizes and types. Responsive design has the potential to solve these problems as well as allow user interfaces to be reused on new platforms, such as phones. Responsive design is commonly used for scaling content-rich web pages for different platforms. However, even if the content is different, we may use many of the patterns of responsive design for maritime applications.

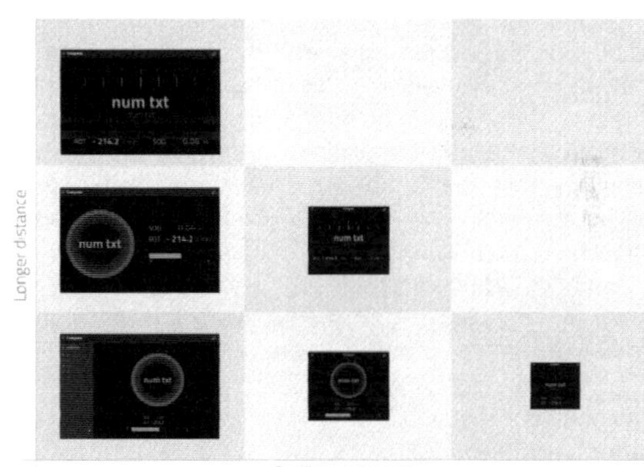

Figure 8: Responsive scaling of size and distance.

Our concept uses a simple scaling scheme in which text, icons and other user interface components maintain a consistent scale when the user interface is scaled for a larger or smaller screen size. In addition, we support independent scaling of text and interaction components based on the user's distance from the screen (Figure 8). This is particularly important on ships where screen distance may vary tremendously.

In order to keep interaction components from overlapping at small screen sizes, we have introduced two breakpoints. A breakpoint defines a certain size threshold where the user interface is rearranged to better facilitate a new range of display area sizes. In our design, the thresholds rearrange the menu structure in the interface. The first collapses the right action menu, while the second removes all remaining menus and effectively transforms the application to a widget-type app.

Our trials show that we can replicate most of our applications with responsive design to allow them to scale freely from very small form factors to large screens. Adjusting to smaller form factors requires the hiding of parts of the interface, and many of the applications will need a defined minimum screen area to ensure that they can safely show essential content for maritime operations.

There are multiple reasons for adopting responsive design in maritime applications. The modern ship's bridge already has multiple screen types and sizes. Moving forward, we expect there will be additional formats made available, including higher screen resolutions, larger screens and portable formats. Further, as systems are accessible over a network, it is likely that users will access maritime applications in many places. Finally, these applications are installed on multiple ships, and as available screen formats increase, it is likely that the same application will be shown on different screen formats on different ships.

4. DISCUSSION

The presented concept is a work in progress and not a finished design system. However, based on our experience so far, we argue that the work shows that it is possible to create a maritime design system based on web technology that can achieve consistent user interfaces across a wide variety of applications. The implementation of a maritime design system may realise the goal of design consistency across applications, leading to improved usability when compared with the current MBS implementations we have observed.

Although the main contribution of the concept is to offer a pathway for *inter-MBS consistency* (consistency across several systems) and *intra-MBS consistency* (consistency within a single multivendor ship's bridge system), we believe the system can also improve individual applications. Such a design system would go much further than current maritime regulations in specifying how to design applications for maritime use. This would be especially beneficial for smaller companies with limited resources for user interface design. Many of our partners report that clearer guidelines would improve the design quality of their current offerings as well as reduce design and implementation costs.

The applications we have designed are self-contained individual applications that will sometimes exist within a larger integration system, much like apps on smartphones. This represents a challenge for the designer who may not know how the main integration system organizes menus or switches between applications. Our design system handles this by providing for the consistent placement of global navigation in each application. However, if an integration system design conflicted with our approach, a change would need to be made either to the integration system or to our design system. Examples of such conflicts could be a global top bar in the integration system that obscures the application top bar or flyout menus for navigation in the integration system that overlap application menus. To avoid such conflicts, we argue that a successful maritime design system must include specifications for how integration systems should facilitate the integration of applications. We will extend our design work to encompass requirements for integration systems as we further develop the concept.

Many of our partner companies use interface design to differentiate their products. This is a barrier to introducing a uniform maritime design system. Many of the industry actors we collaborate with are worried that a strong design system would reduce their ability to stand out in the market. We have made efforts to relieve such concerns by opening up the possibility of differentiating user interfaces through styling (facilitated by CSS) while maintaining consistency across MBS through the layout and components. This would make it possible for individual applications or entire bridges to be presented in company-specific styles. However, our design system will still strongly regulate individual actors' ability to freely design large portions of their user interface.

We have found no equivalent system openly available for maritime application development. The only systems we have found are companies' internal design systems that cannot be used by outside companies. There are also significant differences between our design and the current dominant web-based design systems. First, the maritime design system must be very rigid in accordance with our strong emphasis on consistency. Second, use of colours must adapt to maritime needs, providing for day and night palettes and the restriction of specific colours to particular functions. Third, a range of maritime-specific user interface components is required. Fourth, the system must be able to support multiscreen environments; and fifth, the system needs to integrate maritime regulations that affect user interface design.

Because of differences between maritime needs and the objectives of current web-oriented design systems, we cannot directly apply these existing systems to the maritime domain. Instead, we need to adapt these systems to maritime-specific needs. However, by using web-based systems as a basis for new design, we may build on the tremendous progress already made in a very large global industry in order to improve maritime systems.

Our current work focuses on graphical user interfaces. However, maritime user interfaces exist in in shifting and demanding conditions that affect users' ability to make use of equipment [16]. Because of this, we argue that maritime design systems will need to include multimodal and in particular tangible user interfaces [2,16]. We will explore such options in future research.

5. CONCLUSION

We have presented the first stage of a design system for maritime user interface design. Our work shows how web-based design methods and tools may be applied to the maritime industry in order to achieve consistent design in MBS. Our concept provides an example of how we can achieve consistency across layout, user interface components and style. The approach opens up the opportunity for multiple constellations of consistency, where for instance application layouts and use of components are consistent across any MBS using the same design system while style is managed within the individual MBS. Such a maritime design system can be an important contribution to improved user interfaces and UX in MBS.

6. ACKNOWLEDGEMENTS

Thank you to the Openbridge consortium for contributing to this article. The project is funded by the Openbridge Consortium and the Norwegian Research Council.

7. REFERENCES

1. MIŠKOVIĆ, D., T. BIELIĆ, and J. ČULIN, 'Impact of Technology on Safety as Viewed by Ship Operators', *Transactions on Maritime Science*, 2018, **7**(01): pp. 51–58.

2. NORDBY, K. and A.D. MORRISON, 'Designing Calm Technology and Peripheral Interaction for Offshore Service Vessels', *Personal and Ubiquitous Computing,* 2016, 20(4): pp. 601–613.

3. OPENBRIDGE, 'Openbridge Project Webpage', 2018 [cited 2018 April]; available from: www.openbridge.no.

4. NIELSEN, J., 'Coordinating User Interfaces for Consistency', 2014: Elsevier.

5. GRUDIN, J., 'The Case against User Interface Consistency', *Communications of the ACM,* 1989, **32**(10): pp. 1164–1173.

6. HAGEN, J.E., 'Implementing e-Navigation', 2017: Artech House.

7. APPLE, 'iOS Design Themes' [cited 2018 April]; available from: https://developer.apple.com/design/human-interface-guidelines/ios/overview/themes/

8. GOOGLE, 'Material Design' [cited 2018 April]; available from: https://material.io/guidelines/

9. MICROSOFT, 'Design Basics for UWP Apps', [cited 2018 April]; available from: https://docs.microsoft.com/en-us/windows/uwp/design/basics/

10. APPLE, 'macOS Design Themes', [cited 2018 April]; available from: https://developer.apple.com/design/human-interface-guidelines/macos/overview/themes/

11. CURTIS, N., 'Modular Web Design: Creating Reusable Components for User Experience Design and Documentation', 2010: New Riders.

12. BEVAN, N. 'Classifying and Selecting UX and Usability Measures', *International Workshop on Meaningful Measures: Valid Useful User Experience Measurement*, 2008.

13. LAW, E.L.-C., et al., 'Understanding, Scoping and Defining User Experience: A Survey Approach', *Proceedings of the SIGCHI Conference on Human Factors in Computing Systems,* 2009: ACM.

14. MALLAM, S.C. and K. NORDBY, 'Assessment of Current Maritime Bridge Design Regulations and Guidance', 2018, The Oslo School of Architecture and Design: Oslo, Norway.

15. GARDNER, B.S., 'Responsive Web Design: Enriching the User Experience', *Sigma Journal: Inside the Digital Ecosystem,* 2011, 11(1): pp. 13–19.

16. NORDBY, K. and A.D. MORRISON, 'Multimodal Interaction for Marine Workplaces Used as Strategy to Limit Effect of Situational Impairment in Demanding Maritime Operations', *International Conference on Marine Design*, 2015: London.

8. AUTHORS BIOGRAPHY

Kjetil Nordby holds the current position of associate professor at The Oslo School of Architecture and Design. He is research manager for Ocean Industries Concept Lab and project leader of the Openbridge project. His previous experience includes interaction design, industrial design, design management and research management within the ocean industries.

Synne Frydenberg has been a PhD fellow since 2017 at The Oslo School of Architecture and Design. She also works part time as a design researcher within the SEDNA project. Her research interests lie in how to design for situated interaction in complex professional settings, such as a ship's bridge. Before starting her PhD, she worked six years as an interaction designer in the Norwegian design consultancy Making Waves and in the Norwegian publishing house Gyldendal's digital department.

Jon Fauske holds the current position of research assistant at The Oslo School of Architecture and Design. He is responsible for user interface design of the Openbridge design system. His previous experience includes industrial and interaction design.

PRESENTATION

A HARMONIZED DISPLAY OF E-NAVIGATION INFORMATION; WORKLOAD AND NAVIGATIONAL CONTROL; CONTEXTUAL CONTROL LEVELS

T Porathe, Norwegian University of Science and Technology, Norway
M Lutzhoft, Western Norway University of Applied Sciences, Norway

SUMMARY

As a result of the work with the e-navigation initiative, IMO is now working on "a harmonized display of navigation information received via communications equipment". The availability of a multitude of new important, but possibly complex, information on the ship bridge will be a challenge for human factors and design as the risk for information overload is obvious. This paper proposes a new framework for structuring the design of the information environment on the ship bridge. The paper starts with a brief reference to some maritime accidents where a mismatch between the availability of information and human performance can be detected. Using Hollnagel's theoretical constructs of information processing and cognitive engineering we suggest that Hollnagel's Contextual Control Model theory can be used to structure equipment and bridge layout based on operator workload. The intention is to advance HMI design and integration of bridge equipment based on different workload levels.

1. INTRODUCTION

Shipping is an industry with longstanding traditions. Previously, information available on the deck of a sailing vessel or on a steam ship bridge, was limited to a few coarse instruments and often only rules of thumb learned through experience. Today we have a plethora of information available. To give just two examples: weather reports have been received by radio for decades and helped ships avoid bad weather, as compared to having only barometers and knowledge of the general paths weather systems take in different regions. Today nautical chart updates are transmitted to ships by satellite connection whenever fairways are changed or new shoals are detected.

And more information will be available. In 2006, the International Maritime Organization (IMO) started working on a digitalization concept termed *e-Navigation* to "harmonize the collection, integration, exchange, presentation and analysis of marine information onboard and ashore by electronic means to enhance berth to berth navigation and related services for safety and security at sea and protection of the marine environment" [1].

e-Navigation not only holds a promise to enhance knowledge onboard and thereby increase the safety and efficiency of navigation, but also to support a sustainable future industry. The new enhanced information flows will be structured through a number of Maritime Service Portfolios. However, the problem of where and how to display all this new information must be taken seriously and IMO has added a new item to its agenda on *the harmonized display of navigation information received via communication equipment.*

1.1 HARMONIZED DISPLAY OF NAVIGATION INFORMATION RECEIVED VIA COMMUNICATION EQUIPMENT

A first draft of these guidelines was submitted to the IMO's sub-committee on Navigation, Communications and Search and Rescue in December 2016 [2]. The IMO's Maritime Safety Committee approved an Interim Guidelines for the Harmonized Display of Navigation Information Received via Communications Equipment in May 2018 [3].

These interim guidelines states "Navigation information received via communication equipment should be displayed in a timely, unambiguous and harmonized manner. The type and level of information displayed should complement the user's capabilities and should take into consideration human factors principles [- - -]. Higher levels of integration mean that systems should be carefully evaluated to ensure that complexity and workload are compatible with the ability of the user (OOW)." (Annex, page 2).

Furthermore the guidelines state "Navigation information should be displayed in such a manner that information overload is prevented. Selectable functions should be included to allow for display of only the required information necessary for safe navigation and the task at hand." (Annex, page 3). And, "It should be possible to select and filter (categorize) information and data received onboard in accordance with urgency and sea area." (Annex, page 3).

The implication here is that information needs to be designed differently depending on the urgency of the decision-making and the workload of the operators. We will now set this statement into context. Let us therefor start out by briefly look at the work situation on the ship bridge.

2. DECISION-MAKING ON THE BRIDGE

Modern oceangoing cargo ships usually have four bridge officers, in total, responsible for the navigation of the ship. The captain has the overall responsibility for the ship and does not take watches under normal circumstances. The other three officers are working on a round-the-clock schedule. During daylight, the watch officer is usually alone on the bridge in open sea conditions. The work consists of monitoring course keeping and the surroundings, looking out for other ships or dangers. This is done by binoculars, chart and radar. Other ships, equipped with an Automatic Identification System (AIS) transponder will be visible on the Electronic Chart and Display Information System (ECDIS) when within radio range. Besides position, course and speed, the AIS will show other information such as name and destination. This is the case for larger ships, but smaller vessels like fishing and leisure crafts might not carry AIS and will only be visible on the radar.

2.1 OCEAN PASSAGE

There are of course a plethora of other tasks such as recording position fixes and documenting the voyage, but generally, during ocean passages, the workload of the watch officer can be low. With no ships around and nothing on the radar screen for 24 nautical miles, the problem might instead be one of boredom (low arousal). During darkness, a seaman is added to the bridge team as a lookout, both because the degraded visibility puts higher demands on vigilance, but also as a safeguard against one or the other falling asleep.

2.2 COASTAL NAVIGATION

As the ship is approaching its destination and entering coastal waters the traffic intensity may increase, buoyage systems need to be found and traffic separations adhered to. This puts higher demands on the human on the bridge. More observations need to be made and more decisions taken during a given time interval. The workload has now increased compared to the open ocean situation.

2.3 CONFINED WATERS

As the ship is approaching port, the workload increases even more. Radio contact needs to be taken with the pilot station and pilots picked up off the coast. During the approach through a complicated archipelago, great attention needs to be given to the charts and other vessels in the area. Communication increases, with various stakeholders in port to prepare for the arrival, and rigorous time keeping is exercised to arrive in time for tugboats and linesmen. Often this increased workload is mitigated by adding more persons to the bridge team, e.g. the maritime pilot and maybe an extra officer.

2.4 ARRIVING IN PORT

Finally, the ship arrives in port. The workload on the bridge now increases even more. The traffic in a busy port may be heavy, a lot of energy needs to be spent on manoeuvring with wind and current now affecting the slow moving vessel. Mooring lines need to be ready and a thousand other preparatory tasks performed. The workload is very high and the captain also joins the bridge team, often doing the final manoeuvrings to berth.

Of course, workload may peak at any time due to unforeseen circumstances, which will be illustrated later on.

The diagram in Figure 1 schematically describes the general workload on the bridge as the quota between resources demanded (both cognitive and physical such as attention and time) and resources supplied. To the left in the figure we can see that the resource demand during an ocean passage is generally low and then increases as the ship enters coastal waters and finally the confined waters of the archipelago and the port. We can also see that for any individual there is a maximum level of resources available, due to e.g. skills and experience. This maximum level may differ between individuals, but may also vary within one individual due to e.g. fatigue or drug use. (The diagram is, of course, only a very general description and may vary for any particular voyage.)

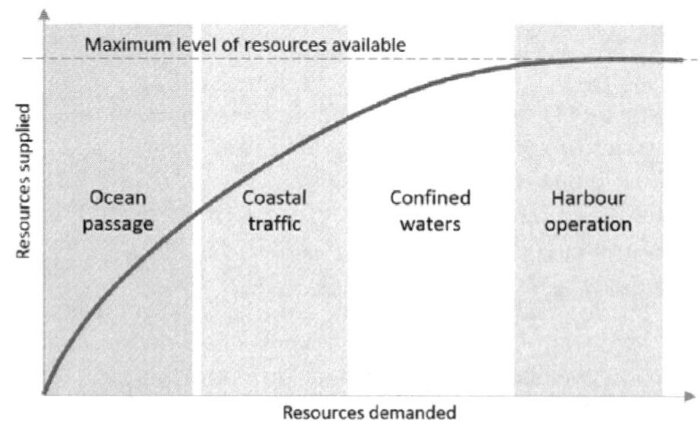

Figure 1: The diagram shows the relationship between resource supply and demand during typical phases of a ship's voyage. Resources are here both cognitive and physical and can be supplied by higher arousal (up to a certain limit), by automation or by adding another operator. Adapted from [4].

In the model above, we can see that the resource demand on the bridge changes during different phases of a prototypical voyage. Changing demands on workload can be dealt with on an individual level by increased or decreased arousal, but may also by adding more personnel to the bridge team. However, in abnormal circumstances the situation will have to be handled with the resources available on the bridge at the time of the incident. We now take a closer look at the human operator.

3. WORKLOAD AND ACCIDENTS

We are in this paper discussing *mental workload*, the subjective experience of mental effort. Just like physical workload, the concept of mental workload has been developed to refer to the attentional demands experienced during the performance of cognitive tasks [e.g. 5, 6, 7].

We will illustrate the workload of bridge officers by briefly telling the story of two ship accidents.

3.1 THE SLEIPNER ACCIDENT

In 1999, the high-speed ferry *Sleipner* crashed into a rock on the Norwegian coast and 16 of 85 of the passengers and crew drowned as the ferry subsequently sank. It was a dark November evening with gale winds and high waves. The ferry was underway doing 35 knots over a stretch of open water. The captain and the first officer were seated in an airplane cockpit arrangement in the wheelhouse and navigating visually using the sector lights of the beacons along the coast. Due to wave induced sway, the autopilot had been disengaged and the captain was steering manually using a turn knob on the armrest of his chair. Approaching *Haarskru* lighthouse, both officers independently turned to their separate radar sets to make adjustments. No one was looking out of the window to find the white sector of the next beacon. After having adjusted his set, the first officer looked up and saw the flash from *Haarskru* lighthouse, but now in red sector, and on the starboard bow. He immediately realized that they were in danger of hitting a rock in the starboard side of the fairway and called out a warning. The captain looked up and turned on the floodlight of the vessel. The rock appeared just ahead of *Sleipner* and the captain gave full port rudder and reversed the engines, but the manoeuvre came too late and the vessel crashed on to the rock in almost full speed [8].

One may in hindsight note that although the ferry had a fully functional, state-of-the-art bridge, with an electronic chart system, the officers did not manage to recover their situational awareness until it was too late. Although all information was available, however, in some case (as the electronic chart), difficult to observe, there was simply not enough time to read all the necessary instruments and integrate them into a conclusive image of the situation. The resources available to the two officers simply did not match the resources demanded (see Figure 1).

3.2 THE RENA ACCIDENT

One night in October 2011 the 225-meter long container vessel *Rena* grounded on the *Astrolabe Reef* off the North Island of New Zeeland. The vessel was severely damaged but remained on the reef for several months before she finally broke up, causing environmental pollution. *Rena* was approaching the pilot pick-up station for Tauranga Port but needed to hurry because she was at the end of the tidal window allowing entry to the port. The second officer decided to take a shortcut south of the planned track bringing the vessel within one nautical mile to the Astrolabe Reef. *Rena* had no electronic chart system and used paper charts placed in the chart room at the back of the bridge. The voyage was monitored by position fixes plotted with pencil on the paper chart in the chart room. As the *Rena* approached Tauranga pilot pick-up station the captain joined the second officer and the lookout on the bridge. The captain noticed an intermittent echo on the radar screen about 2.6 nautical miles dead ahead of the vessel. He and the lookout tried to detect what was causing the echo using binoculars both from the bridge and the bridge wing, but without success. As the captain went to the chart room to plot the ship's position on the chart the *Rena* struck the reef at 17 knots [9].

It is in hindsight interesting to note that part of the information needed to keep up the operator's situation awareness was stored in two different places. The nautical chart, with own ship's positions and the position of the reef, was placed in the chart room at the back of the bridge, while the only real-time information provided, was by the radar available at the front of the bridge. The radar did not reveal the position of the reef (apart from the intermittent echo of breaking waves). The resources available to the two officers and the lookout did not match the resources needed to integrate the three information sources; chart, radar and visual search.

3.3 WORKLOAD AND INFORMATION DISPLAY

From an operator performance perspective the two accidents referred to above are very different. In the *Sleipner* case, the officers suddenly became aware of imminent danger and under the influence of stress and time pressure was not able to recover the situation. In the *Rena* case, the officers were unaware of the impending danger and the workload remained relatively low all the way up to the grounding, although they struggled to solve the mystery of the intermittent echo.

In both the *Sleipner* and the *Rena* cases, the presentation of information did not support solving the problem they were heading into. The *Sleipner* had an electronic chart system showing both the danger and the position of the vessel relative to the danger, but the chart was located in an off-view position and presented in a format that did not support immediate action. On the bridge of *Rena* information was fragmented in two different locations and needed to be "carried in the head" between the chart room and the radar in order to integrate it into a solution that made the problem transparent.

The underlying assumption in both these cases is that the information could have been designed in a better way to support immediate decision-making – a less demanding design. But before we return to *Sleipner* and *Rena* and our final goal of addressing the design of information, we make reference to the theoretical framework of Eric Hollnagel's Contextual control model (COCOM).

4. THE CONTEXTUAL CONTROL MODEL

The Danish psychologist Erik Hollnagel, developed the cyclic Contextual Control Model as a response to the fragmented classical stimulus-response based human-machine view in 1993 [10]. This model depicts human actions, as the consequence of the context they work in and feedback from earlier actions. Their activities are also the result of feedforward expectations based on their current understanding of the situation, thus becoming a spiral of steps in a "joint cognitive system" of man and artefacts. See Figure 3.

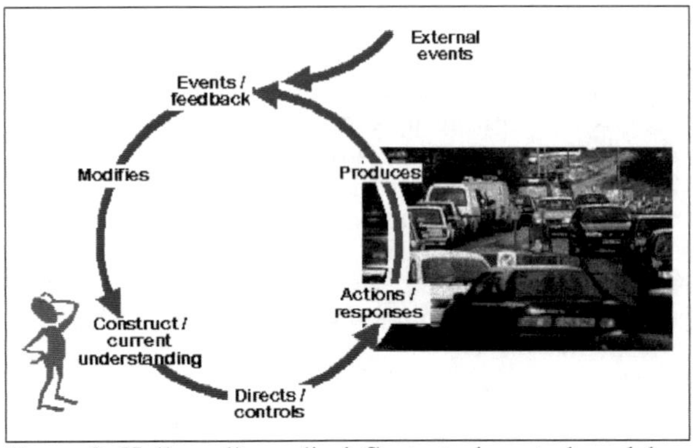

Figure 3: Hollnagel's cyclical Contextual control model (COCOM) [11]

An important feature of COCOM is the control modes, which corresponds to "characteristic differences in the orderliness of performance" [12]. If we define the system to encompass all involved parts of the navigation task of a ship, we can apply the control levels of a joint system. The four control modes Hollnagel suggests are: Strategic, Tactical, Opportunistic and Scrambled. Hollnagel points out that they should be considered as regions on a continuum, which ranges from no control, to completely deterministic performance. Hollnagel describes these control levels as follows:

4.1 STRATEGIC CONTROL MODE

In the strategic control mode, the joint system has a wider time horizon and can look ahead at higher-level goals. The dominant features of the situation or the interface therefore have less influence on the choice of action. At the strategic level the functional dependencies between task steps and the interaction between multiple goals will also be taken into account in planning [12].

4.2 TACTICAL CONTROL MODE

The tactical control mode corresponds to situations where performance more or less follows a known procedure or rule. The joint system's time horizon goes beyond the dominant needs of the present, but planning is of limited scope or range and the needs taken into account may sometimes be ad hoc [12].

4.3 OPPORTUNISTIC CONTROL MODE

In the opportunistic control mode, the salient features of the current context determine the next action. Planning and anticipation is limited, perhaps because the context is not clearly understood or because there is limited time available. Opportunistic control is a heuristic that is applied when the constructs are inadequate, either due to lack of competence, an unusual state of the environment, or detrimental working conditions. The resulting choice of actions is often inefficient, leading to many useless attempts being made [12].

4.4 SCRAMBLED CONTROL MODE

Finally, in the scrambled control mode, for humans there is little, if any, reflection or cognition involved but rather a blind trial-and-error type of performance. This is typically the case when situation assessment is deficient or paralyzed and there accordingly is little or no correspondence between the situation and the actions. The scrambled control mode includes the extreme situation of zero control [12].

In the perspective of the work situation of the officers on the ship bridge, Hollnagel's control model can be used to structure the information environment to fit varying time constraints and workload situations of the operators, different resource demands and stress levels, so that "the type and level of information displayed should complement the user's capabilities and should take into consideration human factors principles" [3] as stated in the IMO document in the introduction of this paper. Such a structure might be helpful for interaction designers when designing the new information environment for e-Navigation during the coming years.

5. CONTEXTUAL CONTROL LEVELS ON THE SHIP'S BRIDGE

What does it mean to move between control levels and how can you a) avoid climbing from a level where you have appropriate control, and b) if you are in scrambled or opportunistic, how can you move 'down'? As a basic premise the modes mean that you have more or less time to consider your actions and perform them, or as Hollnagel puts it: time to think and time to do [13]. How do you 'buy time'? Some possibilities are listed below, whereof some may overlap considerably.

1. Consider the ship and crew as one system [12]
2. Slow down the system
3. Get one more person to the team
4. Acquire more competence (long term or short term)
5. Technical usability
6. Design of workplaces
7. Training of crew and designers
8. Use heuristics or ETTO; efficiency-thoroughness trade-off [14]
9. Train people to muddle through – and accept it

How can we use the four contextual control levels when working with bridge design? And do we already accommodate these four control levels in the bridge design we have today – in the case of *Sleipner* and *Rena* for example? Let us go through the four modes one by one.

5.1 STRATEGIC CONTROL LEVEL

Modern bridge procedures and bridge design often support the Strategic level of control well. Voyage planning is done ahead of departure in the relative calm of the chart room or chart table on the bridge. Here you will find the tables, literature and charts needed for careful planning. Several alternative route choices can be compared without time stress. On a modern bridge, you will find an ECDIS planning station and a computer able to go online and receive information and updates from costal administrations and port authorities. This is also the place where strategic planning can be conducted during long ocean passages. Abundant available time also lessens the demands on information design. If information is not understood the first time, there is time to read or look again, or ask for advice. In the diagram in Figure 4 where the control levels are placed on top of the resource diagram presented earlier, the Strategic control level is typically placed to the left, e.g. during passage planning in port, or during ocean passage, with low resource demands.

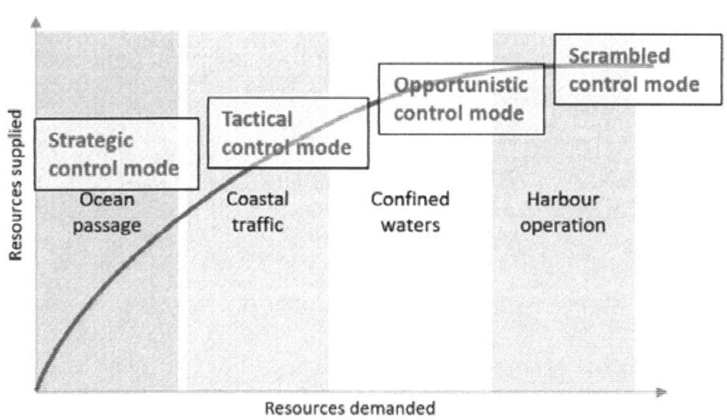

Figure 4: Hollnagel's four contextual control levels overlaid on the diagram in Figure 2, on the different phases of a prototypical voyage.

5.2 TACTICAL CONTROL LEVEL

On a modern ship bridge, you might find an Integrated Navigation System with several screens on which ECDIS, radar and conning information are displayed, as well as other services. If the bridge has a pilot/co-pilot design, you will find two chairs for the watch officer and a co-officer. Although there are several problems with the interaction and information design of radar and ECDIS equipment, they all fairly well support the navigators on the tactical control level, typically illustrated in Figure 4 by the coastal water phase of a voyage. Bridge officers here work with a just-enough-time constraint and need decision support to be clear and unambiguous. Traditionally anti-grounding and anti-collision is taught as two different tasks at the nautical academies, and is also supported by two different instruments, the ECDIS and the radar. The problem here is one of integration, just as it was for the officers on *Rena*. Except that the *Rena* officers did not only need to integrate information from two screens, they had to walk between the chart room and the radar screen situated at different locations.

5.3 OPPORTUNISTIC CONTROL LEVEL

As mentioned, the problem in the *Rena* case was that the position of the submerged reef was not available on the radar screen. Apart from an occasional breaking wave that showed up as an echo. If *Rena* had had a modern integrated chart-radar with the nautical chart as an underlay on the radar screen the solution to the mysterious echoes would have been obvious: the Astrolabe reef depicted by two independent sources, GPS and radar.

"Planning or anticipation are limited, perhaps because the context is not clearly understood or because there is limited time available." (p. 147). Hollnagel writes about the opportunistic control level. Integration of information will be important to support this level, and here much remains to be done. Integrating chart and radar image is a start, already available in modern systems; integrating the right e-Navigation information, without causing clutter or information overload is the next challenge.

5.4 SCRAMBLED CONTROL LEVEL

In the *Sleipner* accident, the two bridge officers suddenly found themselves in a situation where they had temporarily lost their orientation and only a few seconds to regain it. They knew that they were heading for danger but did not know if the danger was to port or to starboard or straight ahead. Nor did they have the time to investigate their position relative to the danger using the ECDIS. In this Scrambled situation, we may conclude that none of the available bridge equipment managed to supply them with the necessary decision support.

Research on navigation and chart displays has demonstrated that cognitive off-loading can be accomplished by removing the need for mental rotations, in turn by presenting the chart view in an egocentric 3D projection [15]. In Figure 5 such a projection is shown on a separate "conning screen" beside the ordinary ECDIS north-up chart view. The ship is southbound and a right direction on the chart translates into a port (left) turn with the ship.

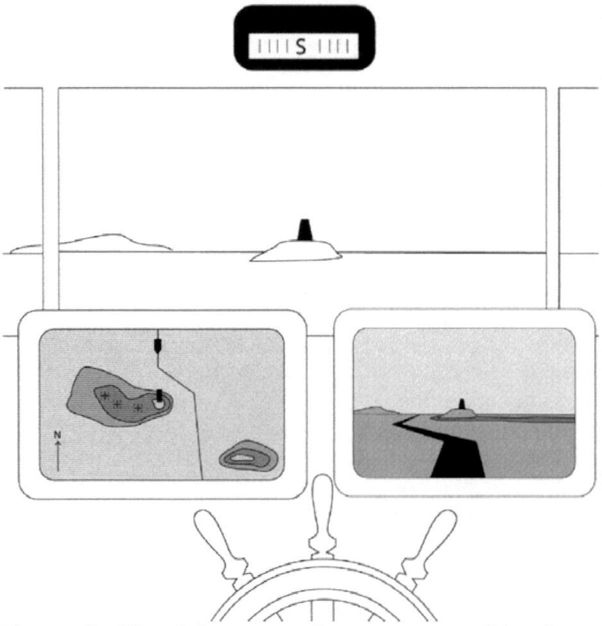

Figure 5: The left screen shows on a traditional exocentric, north-up oriented, nautical chart how the southbound ship is approaching a dangerous rock with shoals on it western side. The helmsman now needs to turn port (left) to follow the planned route, in black. An action that is not always intuitive in time-limited situations due to the mental rotations needed. The right screen shows a suggested "egocentric view" nautical chart, suitable for Scrambled control modes [15].

The cognitive offloading is obtained by showing the chart in an egocentric mode in the right display. The information is prepared and immediately available for quick action, such as which direction to proceed, and if there is enough water under the keel to sail on. Such a projection could also be overlaid on the window as in an airplane head-up display (HUD), or as augmented reality in a pair of glasses. Several research projects are presently underway using augmented reality glasses or visors for maritime use [16, 17].

6. VALIDITY OF COCOM IN NAVIGATION

In an unpublished simulator study in 2008 performed at Chalmers in Sweden, experienced navigators drove ships in different scenarios in the Irish Sea. The purpose was to test differences in navigating with radar and chart in north-up versus head-up modes. One of the parameters recorded was how their subjective "feeling of control changed during the scenario". Before the simulation they were briefly informed about COCOM and the meaning of the four levels of control, and after the scenario run they were asked to draw a graph from the start of the scenario (far left) until the end (far right) on a prepared form and make short notes about situations that changed their sense of control. Figure 6 shows copies of three subjects navigating the approach to the pilot pick-up station in Dublin Bay. The scenario is the same in all three cases but the subjective feeling of control differs (there were some differences in the scenario due to choices made by the subjects). But the point is that no subject had any difficulties to understand the concept of "feeling of control" and although the results are subjective, and different, we argue that the Contextual Control Model could be used as a valid structural framework for workload on the bridge.

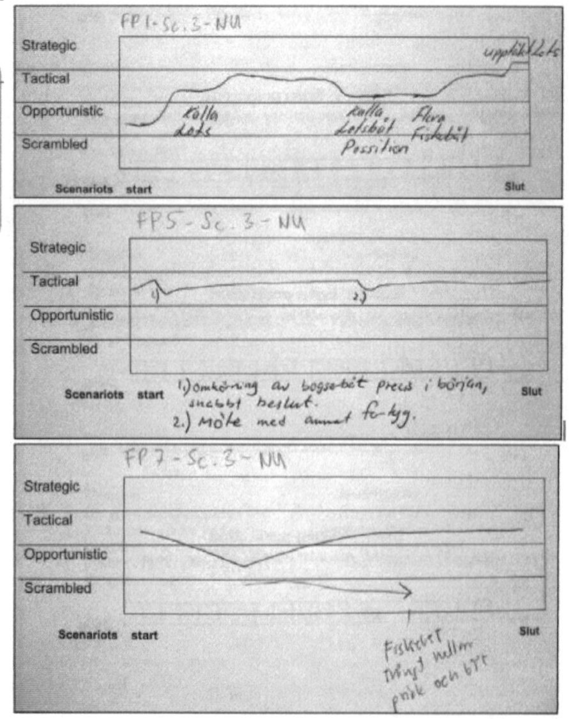

Figure 6: Facsimiles of three response forms drawn by three different subjects after a simulator run at Chalmers in Sweden in the BASSY project in 2008. The scenario took 10 minutes to perform. The data are previously unpublished.

7. DISCUSSION

In the IMO document An Interim Guideline for the Harmonized Display of Navigation Information Received via Communications Equipment, cited in the beginning, we can see that IMO is concerned that the information harvested from new electronic system must be displayed in a way that makes it easy to use for human operators. We are here arguing that the four control levels of COCOM can be used as a framework for assessing where and how on the ship bridge the abundance of new e-Navigation information can and should be presented. We are also arguing that an important facet of the human-machine interface is missing: the information that needs to be pre-integrated and immediately available to support decision-making in "scrambled situations". The right-hand screen in Figure 5 shows an example of such a display where the information needed for the Sleipner officers to make a decision using their "gut feeling" could have been available.

To summarize: In Figure 6 we have made a schematic drawing of a modern ship's bridge. To the right are the two chairs in a pilot, co-pilot bridge set-up, facing right and towards the front of the ship. In this figure, the Scrambled control mode is illustrated by two HUD displays in the windscreen in front of the two operator chairs. To the far left in Figure 6, at the aft part of the bridge, an electronic chart table is illustrating the back-bridge area where the Strategic level voyage planning takes place. Here, we can afford to have a high information density due to an abundance of available time. This is the place where most of the new e-Navigation information should be placed.

In the next section from the left in Figure 6 we have the information displays for the Tactical control level. Here illustrated with the traditional ECDIS and radar screens of a typical modern ship bridge. The separate information on each screen still has to be manually integrated in the head of the bridge officer. He or she is still all right because there is adequate time available in the Tactical control mode. Some new e-Navigation information can be added to the displays without causing information overload. An example is dynamically filtered Maritime Safety Information (MSI), visible directly on the chart in a suitable format when approaching the area in question.

Second to the right in Figure 6 the Opportunistic control level has been illustrated by an integrated radar with chart underlay (or an ECDIS with radar overlay). The integrated display could be in "head-up" mode, meaning the forward-direction of the ship is up on the display, removing one component of the mental rotations needed to translate chart views to actual ship manoeuvres. In some cases, this is cognitively off-loading (e.g. navigation in narrow archipelago fairways), in other situations the traditional north-up orientation is preferable (e.g. communicating with other ships in high traffic density). In this control mode, as well as in the Scrambled mode, the new e-Navigation information needs to be carefully selected and prepared to be useful. This will be a challenge for future research and design.

8. CONCLUSIONS

This paper suggests a way of looking at ship bridge design based on resource demands in different phases of a ship's voyage as well as the workload of the bridge operators. Hollnagel's Contextual Control Model (COCOM) is used as a framework and the strategic, tactical, opportunistic and scrambled control levels identified as useful abstracts for different workload levels on the bridge. The work places of the four different control levels might be at different locations on the bridge, as suggested in Figure 6.

By designing different strategic, tactical, opportunistic and scrambled ship simulator scenarios, different types of new "e-navigation equipment" could possibly be benchmarked and tested. This could prove an interesting method to support usability and human-centered design in a domain where information overload is a risk.

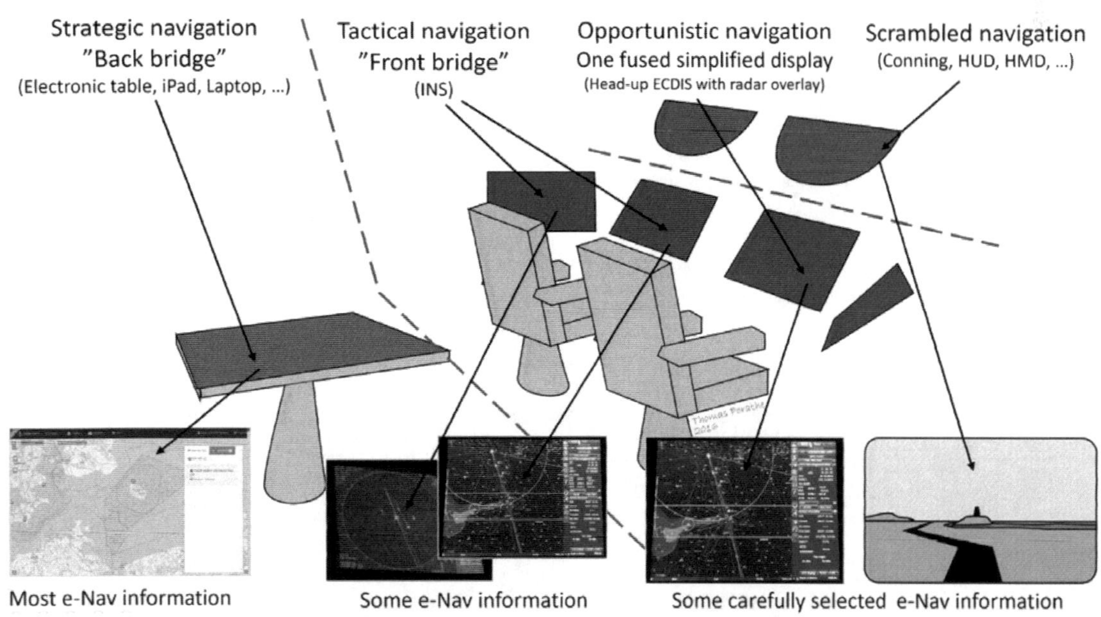

Figure 6: The four different contextual control levels supported by different information displays and HMIs in different locations on the ship bridge (illustration by Porathe).

9. REFERENCES

1. IMO, 'e-Navigation', http://www.imo.org/en/OurWork/Safety/Navigation/Pages/eNavigation.aspx [Acc. 2018-04-08].

2. IMO, 'Guidelines for the harmonized display of navigation information received via communications equipment', NCSR 4/8, 2016.

3. IMO, 'Interim guidelines for the harmonized display of navigation information received via communication equipment', MSC.1/Circ.1593. 2018.

4. WICKENS, C.D., HOLLANDS, J.G., BANBURY S., & PARASURAMAN R., 'Engineering psychology & human performance (4th ed.)', *Psychology Press*, 2012.

5. O'DONNEL, R.D. and EGGEMEIER, F.T., 'Workload assessment methodology'. In K.R. Boff, L. Kaufman & J.P. Thomas (Eds.) *Handbook of human performance. Vol. 2. Cognitive processes and performance.* Chichester: John Wily & Sons Ltd. 1986.

6. PARASURAMAN, R. and HANCOCK, P.A. 'Adaptive control of workload'. In P.A. Hancock & P.E. Desmond (Eds.) *Stress, workload and fatigue* (pp. 305-320). Mahwah, NJ: Erlbaum. 2001.

7. WICKENS, C.D. 'Multiple resources and mental workload'. *Human Factors*, 50, 449-455. 2008.

8. NOU. NORGES OFFENTLIGE UTREDINGER, 'Hurtigbåten MS Sleipners forlis 26. November 1999. Rapport fra undersøkelseskommisjonen oppnevnt av Justis- og politidepartementet 1. desember 1999, NOU 2000: 31', Oslo: Statens forvaltnings-tjeneste, Informasjonsforvaltning. 2000. https://www.regjeringen.no/no/dokumenter/nou-2000-31/id143395/ [Acc. 2018-04-08]

9. TAIC, TRANSPORT ACCIDENT INVESTIGATION COMMISION, 'Containership MV Rena grounding on Astrolabe Reef 5 October 2011', Final Report Marine inquiry 11-204, 2014. https://taic.org.nz/inquiry/mo-2011-204 [Acc. 2018-04-08]

10. E. Hollnagel, 'Models of cognition: procedural prototypes and contextual control', *Le Traviaille humane*, 56, vol. 1, pp 27-51. 1993.

11. E. Hollnagel, "Contextual control model (COCOM)". http://erikhollnagel.com/ideas/cocom.html [Acc. 2018-04-08]

12. E. Hollnagel, and D.D. Woods, 'Joint cognitive systems: foundations of cognitive systems engineering', London: Taylor & Francis, p. 146. 2005.

13. HOLLNAGEL, E. 'Time to think and time to do? I can fail and so can you!', International Australian Aviation Psychology Symposium. 2003.

14. HOLLNAGEL, E. 'The ETTO Principle: Efficiency-Thoroughness Trade-Off: Why Things That Go Right Sometimes Go Wrong', CRC Press. 2009.

15. PORATHE, T., '3-D Nautical Charts and Safe Navigation', Malardalen University Press, 2006.

16. PECOTA S.R., 'The maritime head-up display and high speed vessels: a natural fit?', *Proceedings of the International Symposium. Information on Ships*, ISIS, pp 30-31. 2012.

17. GRABOVSKI M., 'Research on wearable, immersive augmented reality (WIAR): adoption in maritime navigation', *Journal of Navigation*, vol. 68, Issue 3, pp. 453-464. 2015.

10. AUTHORS BIOGRAPHY

Thomas Porathe holds the current position of professor of interaction design at the Norwegian University of Science and Technology in Trondheim. He has for many years worked with design of information systems within e-Navigation and Human Factors and related projects. He is active in the IALA e-Navigation committee.

Dr Margareta Lutzhoft is a Professor in the MarSafe group at the Western Norway University of Applied Sciences. Her research interests include Human Factors, human-centered design and the effects of new technology.

PRESENTATION

The Implementation of e-Navigation Services: Are We Ready?

K Aylward, **R Weber**, **M Lundh**, and **S MacKinnon**, Chalmers University of Technology, Sweden

SUMMARY

Advanced technologies and services are being developed to improve safety and efficiency of navigation at sea. The Sea Traffic Management (STM) EU Project is one of the largest e-Navigation projects examining the digitalization of the shipping sector. The newly developed "STM Services" allow ships to share their monitored route with other ships, planned and/or monitored routes with shore centers and receive suggested routes from shore centers. These tools also allow ships to receive navigational warnings, ETA updates, and chat messages from shore center services. The European Maritime Simulator Network (EMSN), consisting of twelve ship handling simulators based in seven EU countries, examines the potential impact of these services through full scale simulations with scenarios involving up to 30 ships, complimenting the feasibility and understanding of implementing the STM services in live test beds.

Two weeks of full scale EMSN simulations with seafarers were completed comparing navigation as it is today with navigation using STM Services. Structured debriefing sessions were completed at the end of every day, to obtain feedback on the STM Services. The feedback from participant debriefs was analyzed in relation to safety, efficiency, and usability of the STM Services.

The results indicate that the seafarers are positive about the idea of e-Navigation services, and believe that these services could provide additional time to plan, act, and respond to a navigational challenge. However, there were concerns related to their implementation, and the potential negative affect of digitization of current navigational processes which may lead to an increased risk for incidents related to situation awareness, mis-communication, and human error.

1. INTRODUCTION

In 2009, the International Maritime Organization (IMO) formally adopted e-Navigation and published their strategy for developing an implementation plan. The scope of e-Navigation, as defined by the IMO, is "the harmonized collection, integration, exchange, presentation and analysis of marine information on board and ashore by electronic means to enhance berth to berth navigation and related services for safety and security at sea and protection of the marine environment" [1]. One of the key concepts of this implementation plan is that the initiative must be led by user needs, including both seafarers and shore side services. The overall goal of e-Navigation is to improve safety by reducing errors, while also increasing the efficiency of ship operations [2-4].

The concept of e-Navigation and the "connected ship" has dominated many discussions regarding shipping in the past decade. Regulatory bodies, classification societies, nautical academics and shipping companies, have formally addressed e-Navigation in terms of guidance and research efforts [1, 5-7]. The benefits of being connected at all times with instant access to additional information should help increase efficiency, promote information transparency, and improve maritime operations. However, the exponential increases in technology have raised concerns related to potential risks associated with increased complexity on the bridge, within the Vessel Traffic Services (VTS) and within shipping logistics [8-11].

Research on technological changes and the impact on work practices including navigational aids have been on-going for many years [2, 8, 12-15]. This research emphasises the need for crew-centered [2] and end-user involvement in the design of ships and ship systems [14] in order to support safe, efficient and effective operations. This early research focused on; single vessels [16-18], an operator/ bridge team and how they interact with e-Navigation systems [12, 16, 17, 19], and the role of the VTS or shore centers in e-Navigation [8, 20, 21].

Information overload [22], cluttered or scattered information on navigational screens, and/or systems have been cited as potential risks resulting from increased automated systems on the bridge [15, 16, 23, 24]. These risks are not only pertinent to the maritime domain, but are commonly cited within other transportation industries (air, car, train, space)[22, 25-29]. The more information made available to the operator may result in more time with "heads down", which reduces situational awareness related to the task [29-32]. The changes in technology and the effects on working environment have also been researched [20, 33, 34] and the results point out several stressors, especially high work demands on bridge crew[33-36].

As the ability to connect and share information has become more available it is important to understand how this will impact the entire maritime logistics chain, and not just individual components of this complex sociotechnical system [9]. The maritime network and its units often operate in a low-bandwidth environment and are subjected to situations and areas with varying communications capabilities which creates challenges to information sharing [37, 38]. To fully develop and implement IMO's e-Navigation strategy, the challenges, including; the speed of technology development, end-user involvement in design, the extreme environment of shipping, lack of standardization within maritime communications equipment, and security considerations in maritime networks, must be overcome. The International Association of Marine Aids to Navigation and Lighthouse Authorities (IALA) has published a Maritime Radio Communication Plan (MRCP) which focus on a need for "an agreed infrastructure of communication between ship and ashore" [39] and aims at assisting in overcoming the challenges of the maritime sector and be able to manage more information, more often and on a global basis.

1.1 BACKGROUND

Several European Union (EU) projects have studied e-Navigation and the impact of advanced technologies and levels of automation within the maritime industry. EfficienSea aimed at improving maritime safety in the Baltic Sea region through concise and coordinate actions [40]. Following on this work was the EfficienSea 2 project which created and implemented solutions for efficient, safe and sustainable traffic at sea through improved connectivity for ships [41]. MONALISA was another project focused on development, demonstration and dissemination of innovative e-Navigational services to the shipping industry [16, 18]. Succeeding this project was MONALISA2.0 with a vision of making real-time information available to the shipping domain. MONALISA2.0 defined STM and assessed the strengths and weaknesses of the current maritime ship- and transport systems, operations and interactions, and defined a target concept with key performance indicators[42]. A set of operational services were identified and the STM Validation project created technical services that supported these operational services.

The STM Project (2014-EU-TM-0206-S), examines many aspects of digitalization within the shipping sector. The goals of the STM project are to create a safer, more efficient and environmentally friendly maritime sector. STM connects and updates the maritime world (ships, ports, vessel traffic services, service providers, shipping companies) in real time through information exchange and sharing to offer a digital infrastructure for shipping.

As conceptualized in MONALISA2.0 and advanced in the STM Validation Project, the European Maritime Simulator Network (EMSN) was developed. The EMSN consists presently of 12 connected ship handling simulators based in seven EU countries with the possibility to run scenarios with over 30 manned simulated vessels. The EMSN is a unique testbed that enables the introduction and testing of new technologies in complex and large-scale traffic situations, without exposing seafarers to any risks.

The technologies that have been developed in the STM Project are called the "STM-Services". These services include: Route Cross-check, Route Optimisation, Ship to Ship Route Exchange (S2SREX), Rendezvous (RDV) Navigational Warnings, Enhanced Monitoring, Port Call Synchronisation, Port Call Optimisation, Winter Navigation, Importing Pilot Routes and SAR-Search and Rescue (see http://stmvalidation.eu/stm-services/ for detailed information about all services). The proposed benefits of these services include reduced administrative burden, increased safety/situational awareness, and operational efficiency. Although all of these services are available, they were not all tested in the EMSN trials.

For the purpose of this paper, the services evaluated in this study are hereby referred to as the **STM Services** and are limited to: Receiving Navigational Warnings, S2SREX and RDV, Chat Service, Route Cross Check and Enhanced Monitoring. These services are described in sub-section 2.4. The goal was to evaluate how these STM services may affect navigational safety and efficiency.

1.2 AIM OF THE PAPER

The purpose of this research was to assess the impact of the STM Services on the safety of navigation using the EMSN, in full scale simulations with experienced seafarers. New e-Navigation services such as presenting routes from other ships and receiving suggested routes from a service provider such as a Shore Centre on the Electronic Chart and Display Information System (ECDIS) may have an impact on safety and efficiency of navigation. Additionally, from a holistic approach the aim is to understand how the application of STM Services could impact safety in navigation. This study utilized a mixed methods approach to understand this interaction. This paper focuses on the results from participant feedback about the STM Services used in the simulations during debriefing sessions from eight days of data collection over a two-week period.

2. METHODOLOGY

2.1 PARTICIPANT RECRUITMENT

Professional mariners (active or recently active masters, mates, officers and maritime pilots) were recruited as test participants. Participants were recruited through various social media platforms, professional maritime organizations, and convenience sampling.

2.2 PARTICIPANT DEMOGRAPHICS

An electronic demographic survey was completed prior to arrival at the simulation centers. A total of 227 professional mariners, 33 (15%) women and 194 (85%) men participated in this study. The participants were between 20 and 69 years of age. In terms of navigational experience, 1 participant had less than 1 year of experience, the rest of the participants had between 3 to 31 years, and the majority of participants had between 11-20 years of sea-going experience. The current role of participants varied, 68 participants were captains, 18 Pilots, 29 Chief Officers, 57 Deck Officers, 7 VTS Operators, 16 working in educational services, 14 working in the maritime field, and 18 who did not report in the demographic survey. The test participants were of various nationalities and a breakdown of each simulation center and the number of participants from each respective centre is provided in Table 1.

Table 1: Simulator Centres and Number of Participants at each Centre

Simulator Center	Number of Participants
Swedish Maritime Administration (SMA)/ Chalmers University of Technology (CTH)	21
Aboa Mare (AM)	8
Centro Jovellanos (CJ)	19
UPC Barcelona (UPC)	5
Rörvik (SSR)	78
Fraunhofer CML	28
Flensburg University of Applied Sciences (FUAS)	24
Southampton University (WMA)	32
HS Emden-Leer (EL)	12

2.3 RESEARCH ETHICS

Participants were fully informed of the procedures and risks of the experiment and signed electronic and written Informed Consent prior to the start of the simulations. The experiment complies with the requirements of article 28 of the EU General Data Protection Regulation (2016/679) regarding protection for physical persons in the processing of personal data. Each participant was assigned a unique identification number (ID) prior to arrival, which was used for all the questionnaires throughout the study to maintain confidentiality.

2.4 DESCRIPTION OF THE STM SERVICES

Ship-to-ship route exchange: This service provides the navigator with a route segment consisting of the next 7 waypoints of the monitored route of another vessel. Route segments are broadcasted through Automatic Identification System (AIS) and give additional information to the presently available data obtained by radar/ARPA and AIS. Nothing in the S2SREX information exonerates the navigator from applying the International Regulations for Preventing Collisions at Sea (COLREG) [43] and rather than being used in a close quarters situation, the S2SREX may be used as a tactical tool for supporting decision-making and situational awareness at a longer range.

Rendezvous Function: As an integral part of the S2SREX, this function allows the navigator to view where own ship will meet a target ship if both vessels continue on their monitored broadcasted route with the present speed over ground. This function provides route-based Closest Point of Arrival (CPA) and Time to Closest Point of Arrival (TCPA) based on AIS information.

Shore-to-Ship Route Exchange (Receiving route suggestions from shore): This service allows the shore-centre to send a suggested route to the ship, to be reviewed by the bridge team and then either accepted or rejected. This service can be used in various situations, for example if several vessels are warned to avoid a certain area, the shore centre can plan a route based on all available information and directly send this route to the vessel.

Receiving Navigational Warnings: This service provides a notification which overlays a Navigational Warning Message directly on the ECDIS. If the Navigational Warning involves a geographical area to avoid or be aware of, this will be automatically plotted onto the ECDIS so it is visible to the bridge team.

Chat Function: A standalone software similar to other programs (e.g. Skype) which was integrated on the same station as the ECDIS. Text communications with other stations with enabled STM tools such as Shore Centers and ships.

Enhanced Monitoring and Route Cross Check: After having received a ship's monitored route and schedule shore centres will be able to detect if planned schedule is not kept or if ship deviates from monitored route. A shore centre has the ability to receive any planned route and cross check such route against any navigational dangers and if necessary send a route suggestion back to the ship.

2.5 EXPERIMENTAL DESIGN

In order to test the services, subject matter experts developed two simulator scenarios, one in the South Western Baltic and another in the English Channel/Southampton. The scenarios were strategically designed to test the different functionalities of the STM Services. The Baltic Scenario tested the participants in dense, close quarters traffic situations whereas the English Channel scenario was generally less busy. Each geographical area also had a respective shore center, one located in Southampton, UK and the other in Gothenburg, Sweden. The shore center functioned as a typical VTS center with additional access to the STM services.

2.5.1 DATA COLLECTION

Data were collected over eight days during two non-consecutive weeks in 2018. Nine simulation centres and up to 30 ship bridges in the EMSN were used for data collection. Each simulation centre held a standardized briefing session at the beginning of the day for the participants. The briefing consisted of information about the STM/EMSN, the STM Services used (including step-by-step instructional videos on how to use the services), schedule for the day, and an opportunity to ask questions. The participants were assigned to a two-person bridge team, one Master and one Officer of the Watch (OOW).

After the briefing, a 1.5-hour familiarization session was held on the simulator bridges. Once familiarization was complete, the first exercise began and lasted for 1.5 hours. The first exercise was always the English Channel Scenario as it was designed to be less busy and provided the participants time to explore the usage of the STM services.

Once the simulation was concluded, the participants completed a "post-scenario questionnaire" which asked questions specific to the STM services. The same process was completed in the afternoon for the Southern Baltic Scenario. The test day concluded with a common de-brief with all participants in attendance.

2.5.2 DATA COLLECTION AND ANALYSES METHODOLOGIES

This research utilized a qualitative approach to data collection [44-46] through structured open-ended group debrief sessions at the end of each test day. A between-subjects design was prescribed, however in some cases this was not possible due to participant availability.

The following data were collected throughout the project:

- Background Questionnaire with demographic information
- Post-Scenario Questionnaire (morning and afternoon)
- End-of-day Debriefs: Collective discussions with the participants at the end of the test day

The purpose of the group debriefs [45] was for the test participants to provide an account of their experiences with the STM services, both positive and negative in a non-bias environment. The structure for the debriefs was consistent for all simulation centers. The participants were also encouraged to explain the motivation of their response. The list of questions included:

1. In general, do you think the STM functions/services affect SAFETY in Navigation?
2. In general, do you think the STM functions/services affect EFFICIENCY in Navigation?
3. Do you think that the new STM functions/services will change today's way of working/navigation/management procedures?
4. What training (technical/pedagogical) would be necessary in order to use the new STM functions/services?
5. Do you think Shipping Companies will adopt these tools for their fleet?

A grounded theory approach [44, 46] was used to collect and analyse the qualitative data from the debrief sessions. Software program MAXQDA12 (Release 12.3.5, distribution by VERBI Software GmbH, Berlin Germany) a qualitative and mixed methods software program was used to organize, visualize and analyse the open-ended debrief responses. The coding process was continuous as new themes and relationships emerged between data, this is also known as axial coding [44].

Once it was established that the data were saturated [44], memos explaining the phenomena were recorded. The memos were written to help understand and support the observed findings which are discussed in Section 3.

3. RESULTS AND DISCUSSION

The analysis of the debriefs revealed two common categories which were generally applicable to all STM services. The first common theme was that the STM Services provided seafarers with additional time to plan and respond to emerging navigational situations. This category materialised based on the *positive comments* of each service and coded segments including: planning ahead, situational awareness, trust, automated systems, workload, cost, decision making, and communication. This category encompasses both safety and efficiency, which will be explored in relation to each STM Service in the section below.

The second common category emerged from the negative comments regarding the STM services. Participants generally agreed that the usability of the services caused frustration, which negatively impacted their overall opinion of the services. This category emerged within the coded *negative comments* of each service, and the recommendations for improvement from the participants. The results are described and discussed below.

3.1 PARTICIPTANT FEEDBACK ON STM SERVICES

A summary of the positive and negative comments of each STM service is provided in Figure 1, followed by a detailed analysis of each STM service.

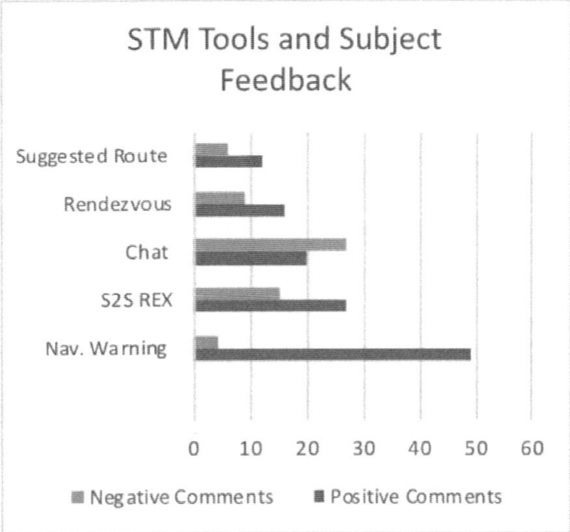

Figure 1: Summary of Positive and Negative Feedback of each STM Service

3.1.1 SAFETY, EFFICIENCY AND INCREASED TIME TO PLAN, RESPOND AND ACT

Understanding the impact of safety and efficiency that the STM services have on navigation practices was the aim of this research. Generally, the participants agreed that the STM services will increase safety and efficiency by providing the operators additional time to plan, act, and respond to events through the automated aspects of the STM systems. Each of the STM services has the potential to reduce the workload for the bridge team, allowing additional time to perform other tasks, and maintain watch of surrounding traffic. Based on the analysis of the participants comments, approximately 60-70% of participants thought that one or all of the STM Services would increase safety and efficiency in some way, approximately 20% said they were not sure/hard to tell because the services were still novel, and 10-15% of participants though that the services could decrease efficiency. A detailed analysis of each STM service is provided below.

Navigational Warning: The navigational warning service was the most highly ranked in terms of general positive feedback with regards to safety and efficiency. A direct quote from a participant was that the service was, "fantastic, and brilliant".

Positives Comments:

- The automatic plotting of the navigational warning could reduce workload and allow the bridge team to maintain situational awareness without the distraction of having to listen, record, and plot coordinates. Participants were able to use that time to perform other navigational tasks.
- The message notification which appeared directly on the ECDIS will alert the bridge team much sooner than a printed Navtex message, which participants said could remain unseen for hours. The instant notification could provide additional time for the bridge team to react to the message if necessary.
- Participants think that this service could reduce human errors
- They believe that this service will reduce "noise in communications" on the Very High Frequency Radio (VHF), as navigational warnings promulgated by VHF could be superseded by the Navigational Warning service and graphically depicted on the ECDIS.

Negative Comments:

- The usability of this service was the only weakness identified, because it was difficult to see the message notification on the ECDIS screen as it was the same color as the menu tab.

S2SREX: Approximately two thirds of participants believe that this service will have a positive effect on navigation, and was coined as a "nice to have" service. However, S2SREX was also controversial because participants are not convinced it will be used properly.

Positive Comments:

- Informative tool that can help with planning and decision making
- This tool allowed ships to identify others intentions and act accordingly.
- It reduced unnecessary communication over the VHF and will reduce misunderstandings. This service replaces the common "what are your intentions?" inquiry over the VHF.

Negative Comments:

- Participants were concerned that ships were not following their broadcasted route which could be a risk if a navigational decision is made based on false information.
- The usability of the tool was a common complaint among participants. The screen became very cluttered when several ships routes were shown which made it difficult to see which route belonged to which ship.
- This service was not considered to be usable in dense traffic conditions with many close quarters situations (e.g. Southern Baltic).
- Over reliability on this service could be dangerous

RDV: This service and the S2SREX service had many overlapping comments. Participants in general liked the RDV service, although there were many issues with the user interface.

Positive Comments:

- This service could increase situational awareness and allow navigators to identify actions of vessels around them and act accordingly.

- This tool offered a helpful visualization of CPA.

Negative Comments:

- Operators must ensure not to over rely on the RDV service. It is a service and it must be used as such.
- Usability was a major issue with the RDV service and many participants agreed that the screen was cluttered which was distracting when the RDV service was being used for several ships at a time.

Chat: This service had the most negative comments related to safety, efficiency, and usability. There may have been a generational gap in the responses to this question. Although this has yet to be analyzed, it appeared that the younger seafarers saw more potential benefit from the chat service than the older seafarers. However, in general the results still showed a negative trend in opinion, and the service was even referred to as "miserable to use".

Positive Comments:

- Potentially useful for ship to shore, and could render less miss-understandings and a clearer information exchange.
- Group chat would be desirable
- This service could increase efficiency if the usability was improved.
- The chat service could help with the issue of English as a second language for many mariners.

Negative Comments:

- Chat messages will distract the operator and draw too much attention to the ECDIS screen, which is a problem already present with ships officers and this should not be encouraged. This service will decrease safety.
- Poor user interface
- It is faster to call a ship on VHF than write a chat message
- Usable for long term situations, but not close quarters
- There is a risk that because the information exchange will only involve two parties (ship-to-ship or shore-to-ship), the rest of the vessels in the area will be unaware of ongoing traffic situations and agreements made through the chat service instead of the VHF.

Suggested Route from Shore Centre: This service wasn´t used as much as anticipated throughout the simulations. However, based on the feedback received, participants generally had more positive comments about the service than negative. This tool was used mostly to test the STM services and not because the participants needed a new route.

Positive Comments:

- Participants were pleased to receive a route when the oil spill was announced in the Baltic Scenario. Many participants were sent a new route, they checked it, and used it.
- Local knowledge at the shore center will improve route quality and thus also operational efficiency

Negative Comments:

- Workload could increase for the VTS Operator, also qualifications / skill set of the VTS Operator could change.
- The captain has to check and approve the new route, which could potentially decrease efficiency and increase workload.
- Liability concerns in case of an incident.

3.1.2 USABILITY OF THE STM SERVICES

There were many negative comments about the usability difficulties of the ECDIS/STM client interface. This is likely expected because of the limitations in design specifications for these types of interfaces and could, in some circumstance, be due to a lack of familiarisation with the experimental equipment. This is one of the most challenging aspects of introducing new technologies in any work environment [15]. The participants in this study are the end user group, which is the focus of IMO´s e-Navigation strategy [1], thus it is critical that these informants be included in product development from the initial design stages[12, 14].

However, despite previous research on Human Centred Design in shipping which clearly show the benefits of early user involvement in design and the importance of usability testing [3, 4, 15], the results of this study confirm that this is still an unresolved issue. The crew did however feel, despite a need for better user interfaces, that the services freed up time, which could have a positive effect on the identified stressful situation on the bridge due to high workload [16, 33-35, 47].

3.1.3 TRAINING, COSTS, COMMUNICATION AND WORKLOAD

In addition to the overarching categories within this analysis - training, costs, communication and workload emerged as sub-categories within the data. The key concepts highlighted by participants are discussed below.

Training

A core aspect of the e-Navigation implementation plan is the educational and training aspect of new services or technologies [1]. Almost all participants agreed that training for the STM Services could be built into an ECDIS model course or an 8-hour training, and would not require an additional stand-alone course. Generally, participants also agreed that a simulation/practical aspect to the training should be required where they can practice using the services in an operational environment. These results align with previous research on maritime training and the importance of a performance-based tasks and the practical component of learning [6, 48, 49].

Cost and Implementation of Services
Participants believe that shipping companies will only adopt the STM Services if they generate an economical improvement, decrease operational costs, or are forced to adopt the services due to regulatory requirements. The general consensus was that shipping companies have little interest in spending money unless it is required by legislation.

Communication
Communication was commonly coded within the data, both within a positive and negative context of STM service integration. The STM services provide an alternative means of communication which is safety critical since decision-making is based on communication between ship-to-ship and ship-to-shore operators and local information integration for preparation and prediction [8, 9].

Communication was most commonly coded within discussions related to the chat function. Some participants believed that agreements made through chat messages between two vessels could lead to reduced situational awareness and increased risk. For example, if a ship made an agreement through chat messages, which would traditionally be made over the VHF, then no other vessel in the vicinity would be aware of this agreement. Participants thought this could negatively impact navigation practices and awareness of ongoing traffic situations and agreements.

On the contrary, miss-communication over the VHF was discussed since English is a second or even third language for many seafarers. A considerable amount of information was mis-understood over the VHF throughout the study and some participants thought that the chat function could improve this mis-communication because the message is written and able to be reviewed whenever necessary. In this case, if communication could become clearer and mis-communications could be decreased then safety would be increased.

Workload
As defined by the IMO, "the VTS are shore-side systems which range from the provision of simple information messages to ships, such as position of other traffic or meteorological hazard warnings, to extensive management of traffic within a port or waterway". The results from the de-brief indicated a potential shift of workload from the bridge crew to the VTS/ shore center which aligns with previous research on e-Navigation [8, 21, 35, 50]. Moreover; manning of shore centers, training requirements for VTS operators, skill sets, and liability issues [8, 21] were also discussed.

The IMO has clear legislation for traditional navigation, which states that while the VTS Operators and pilots may provide navigational assistance, the Master of the vessel is responsible for the safety of navigation [50]. However, the introduction of e-Navigation services challenges the traditional hierarchal structure of shipping. E-Navigation is high priority for the IMO and therefore the uncertainty surrounding liability and regulatory implications will hopefully be resolved in the near future. There is still much validation testing of the available services yet to be completed to ensure the technology is matched with the user's abilities and needs.

4. LIMITATIONS

There are several limitations to consider in this research:

- All the bridges were not always manned with two crew members.
- Senior students were used as subjects in cases when there were not enough certified/qualified participants.

5. CONCLUSIONS

This project highlighted the many benefits, challenges and risks associated with the implementation of e-Navigation services from the point of view of experienced seafarers. The research presented in this paper suggests that in general the STM Services promote navigational safety and efficiency.

These services have the potential to improve communication, decrease bridge crew workload, and increase the time to respond, plan and act accordingly in challenging navigational and traffic conditions. However, rigorous user interface testing is needed to realize the full potential of the services. Additional testing is also required to understand if there is a shift of workload to other aspects of the maritime chain.

Seafarers are generally eager and supportive for the further development and implementation of STM services as long as proper training is provided, and shipping companies are willing to put safety above costs. Even in the earliest stages of development of the STM Services, the participants recognized the potential benefits.

Further testing of the STM and similar services is needed to ensure safety of crewmembers is prioritized. The EMSN and live test beds allow a safe place to test and validate this transition to an e-Navigation based shipping industry.

6. ACKNOWLEDGEMENTS

The authors would like to thank the STM Validation Project (2014-EU-TM-0206-S), the EU commission and co-funding agencies, Connecting Europe Facility/

Motorways of the Sea, Västra Götalandregionen and VINNOVA for financial support. The authors would also like to thank the simulation centres and the participants for their time and willingness to provide detailed feedback about their experiences.

7. REFERENCES

1. International Maritime Organization, *The e-navigation Strategy Implementation Plan (SIP)*, in *MSC 94*, IMO, Editor. 2014: London, UK.
2. van de Merwe, F., N. Kähler, and P. Securius, *Crew-centred Design of Ships – The CyClaDes Project.* Transportation Research Procedia, 2016. **14**: p. 1611-1620.
3. Costa, N., *Human Centred Design for Maritime Safety: A User Perspective on the Benefits and Success Factors of User Participation in the Design of Ships and Ship Systems*. 2016.
4. Costa, N.A., E. Holder, and S.N. MacKinnon, *Implementing human centred design in the context of a graphical user interface redesign for ship manoeuvring.* International Journal of Human-Computer Studies, 2017. **100**: p. 55-65.
5. DNV GL, *The Future of Shipping* 2014, DNV GL: Høvik.
6. Nicolae, F., et al. *THE IMPROVEMENT OF ACADEMIC NAUTICAL EDUCATION BASED ON E-LEARNING TECHNIQUES*. 2013. Bucharest: "Carol I" National Defence University.
7. ABS. *FUTURE READY: ADVANCING FROM AUTOMATED TO AUTONOMOUS SHIPPING* 2018 [cited 2018 15 August]; Available from: https://ww2.eagle.org/en/innovation-and-technology/digital/autonomy.html.
8. Costa, N.A., M. Lundh, and S.N. MacKinnon, *Non-technical communication factors at the Vessel Traffic Services.* Cognition, Technology & Work, 2018. **20**(1): p. 63-72.
9. de Vries, L., *Work as Done? Understanding the Practice of Sociotechnical Work in the Maritime Domain.* Journal of Cognitive Engineering and Decision Making, 2017. **11**(3): p. 270-295.
10. van Breda, L., *Capability prediction: An effective way to improve navigational performance.* Journal of Navigation, 2000. **53**(2): p. 343-354.
11. van Breda, L. and P.O. Passenier, *Effect of path prediction on navigational performance.* Journal of Navigation, 1998. **51**(2): p. 216-228.
12. Lützhöft, M., *"The technology is great when it works" : Maritime Technology and Human Integration on the Ship's Bridge*, in *Linköping Studies in Science and Technology. Dissertations*. 2004, Linköping University Electronic Press: Linköping. p. 108.
13. Aichhorn, K., et al., *Maritime Volumetric Navigation System*, in *Proceedings of the 25th International Technical Meeting of the Satellite Division of the Institute of Navigation*. 2012. p. 3651-3657.
14. Mallam, S.C. and M. Lundh, *The physical work environment and end-user requirements: Investigating marine engineering officers' operational demands and ship design.* Work-a Journal of Prevention Assessment & Rehabilitation, 2016. **54**(4): p. 989-1000.
15. Man, Y., et al., *Gaps Between Users and Designers: A Usability Study About a Tablet-Based Application Used on Ship Bridges*. 2018. 213-224.
16. Lützhöft, M.H. and S.W.A. Dekker, *On Your Watch: Automation on the Bridge.* Journal of Navigation, 2002. **55**(1): p. 83-96.
17. Margareta Lützhöft, J.N., *Integration work on the ship's bridge.* Journal of Maritime research, 2008. **5**(2): p. 59-74.
18. Swedish Maritime Authority. *http://www.sjofartsverket.se/en/MonaLisa/*. 2018 [cited 2018 August 18]; Available from: http://www.sjofartsverket.se/en/MonaLisa/.
19. Praetorius, G., et al., *Increased awareness for maritime human factors through e-learning in crew-centered design*, in *6th International Conference on Applied Human Factors and Ergonomics*, T. Ahram, W. Karwowski, and D. Schmorrow, Editors. 2015. p. 2824-2831.
20. Mansson, J.T., M. Lutzhoft, and B. Brooks, *Joint Activity in the Maritime Traffic System: Perceptions of Ship Masters, Maritime Pilots, Tug Masters, and Vessel Traffic Service Operators.* Journal of Navigation, 2017. **70**(3): p. 547-560.
21. Brodje, A., et al., *Exploring non-technical miscommunication in vessel traffic service operation.* Cognition, Technology & Work, 2013. **15**(3): p. 347-357.
22. Wahlstrom, M., et al., *Human factors challenges in unmanned ship operations - insights from other domains*, in *6th International Conference on Applied Human Factors and Ergonomics*, T. Ahram, W. Karwowski, and D. Schmorrow, Editors. 2015. p. 1038-1045.
23. Vagia, M., A.A. Transeth, and S.A. Fjerdingen, *A literature review on the levels of automation during the years. What are the different taxonomies that have been proposed?* APPLIED ERGONOMICS, 2016. **53**: p. 190-202.
24. Endsley, M.R., *From Here to Autonomy:Lessons Learned From Human–Automation Research.* Human Factors, 2017. **59**(1): p. 5-27.
25. Barham, P., et al., *Jaguar Cars' near infrared Night Vision System - Overview of human factors research to date.* Vision in Vehicles - Vii, ed. A.G. Gale. 1999. 177-185.
26. Batteau, A.W., *The anthropology of aviation and flight safety.* Human Organization, 2001. **60**(3): p. 201-211.
27. Gabbard, J.L., G.M. Fitch, and H. Kim, *Behind the Glass: Driver Challenges and Opportunities for AR Automotive Applications.* Proceedings of the Ieee, 2014. **102**(2): p. 124-136.
28. Turan, O., et al., *Can we learn from aviation: safety enhancements in transport by achieving human orientated resilient shipping environment*, in *Transport*

Research Arena Tra2016, L. Rafalski and A. Zofka, Editors. 2016. p. 1669-1678.
29. Zemrowski, K.M. and Ieee, *Impacts of increasing reliance on automation in air traffic control systems*. 2008 2nd Annual Ieee Systems Conference. 2008. 435-440.
30. Ozyurt, E., F. Flemisch, and B. Doring, *A cooperative automation concept for user-oriented support of air target identification*, in *6th International Conference on Applied Human Factors and Ergonomics*, T. Ahram, W. Karwowski, and D. Schmorrow, Editors. 2015. p. 1918-1925.
31. Han, D.F. and S. Ding. *Review of human factors in maritime system*. Zurich: Trans Tech Publications Ltd.
32. Chauvin, C., et al., *Human and organisational factors in maritime accidents: analysis of collisions at sea using the HFACS*. Accid Anal Prev, 2013. **59**: p. 26-37.
33. Oldenburg, M., B. Hogan, and H.-J. Jensen, *Systematic review of maritime field studies about stress and strain in seafaring*. International Archives of Occupational and Environmental Health, 2013. **86**(1): p. 1-15.
34. Oldenburg, M., et al., *Seafaring stressors aboard merchant and passenger ships*. International Journal of Public Health, 2009. **54**(2): p. 96-105.
35. Boer, L.C., *Workload-watch as an element of human engineering testing and evaluation*. Eleventh Ship Control Systems Symposium, Vol 2, ed. P.A. Wilson. 1997. 105-111.
36. Hetherington, C., R. Flin, and K. Mearns, *Safety in shipping: the human element*. J Safety Res, 2006. **37**(4): p. 401-11.
37. Manoufali, M., et al. *Technologies and networks supporting maritime wireless mesh communications*. in *6th Joint IFIP Wireless and Mobile Networking Conference (WMNC)*. 2013.
38. Kidston, D. and T. Kunz, *Challenges and opportunities in managing maritime networks*. IEEE Communications Magazine, 2008. **46**(10): p. 162-168.
39. International Association of Marine Aids to Navigation and Lighthouse Authorities. *MARITIME RADIO COMMUNICATIONS PLAN (MRCP)*. 2009, revised 2017 [cited 2018 17 August]; Available from: http://www.iala-aism.org/product/maritime-radio-communications-plan/.
40. Danish Maritime Authority. *http://www.efficiensea.org/* 2018 [cited 2018 August 22]; Available from: http://www.efficiensea.org/.
41. Danish Maritime Authority. *https://efficiensea2.org/*. 2018 [cited 2018 August 22]; Available from: https://efficiensea2.org/.
42. Sea Traffic Management. *About Sea Traffic Management* 2018 [cited 2018 20 August]; Available from: http://stmvalidation.eu/about-stm/
43. International Maritime Organization, *Convention on the International Regulations for Preventing Collisions at Sea, 1972 (COLREGs)*. 1972, IMO http://www.imo.org/en/About/conventions/listofconventions/pages/colreg.aspx
44. Corbin, J.M. and A.L. Strauss, *Basics of qualitative research : techniques and procedures for developing grounded theory*. 2015.
45. Patton, M.Q., *Qualitative interviewing*. Qualitative research and evaluation methods, 2002. **3**: p. 344-347.
46. Maxwell, J.A., *Qualitative Research Design: An Interactive Approach (Second Edition)* 2005, Thousand Oaks, California SAGE Publications.
47. Lundh, M. and L.W. Rydstedt, *A static organization in a dynamic context - A qualitative study of changes in working conditions for Swedish engine officers*. Applied Ergonomics, 2016. **55**: p. 1-7.
48. Ghosh, S., et al., *Authentic assessment in seafarer education: using literature review to investigate its validity and reliability through rubrics*. Wmu Journal of Maritime Affairs, 2016. **15**(2): p. 317-336.
49. Ghosh, S., *Can authentic assessment find its place in seafarer education and training?* Australian Journal of Maritime & Ocean Affairs, 2017. **9**(4): p. 213-226.
50. International Maritime Organization (IMO), *IMO Resolution A. 857(20) guidelines for vessel traffic services.*, IMO, Editor. 1997: London, UK: .

8. AUTHORS BIOGRAPHY

Katie Aylward is a PhD student at Chalmers University of Technology in the Maritime Human Factors Research Unit. Her research focuses on human factors, workplace design and advanced technologies in shipping and navigation. Prior to Chalmers, Katie worked as a Human Factors Specialist at the American Bureau of Shipping (ABS) in Houston, Texas. She also holds a Master's of Science in Kinesiology from Memorial University of Newfoundland (MUN), Canada.

Captain Reto Weber, MSc holds the current position of lecturer at Chalmers. He is responsible for the Full Mission Bridge Simulator at Chalmers and head lecturer in ship handling and navigation in confined waters. Reto Weber has sailed in all deck ranks from deckhand to captain on bulk carriers and Ro-Ro vessels. Before joining Chalmers in 2010, he was a maritime pilot on the Swedish West coast. He gained his education in Switzerland, England and Sweden and is holder of a Master Mariner's license (unlimited) and an MSc degree in Maritime Affairs from the World Maritime University. During the last years, Reto Weber has been mainly involved as a simulation manager or project manager in EU projects such as HORIZON fatigue research study, EFFICIENSEA 1 and 2, MONALISA 1 and 2, ACCSEAS and the STM Validation project.

Monica Lundh is currently Head of the Maritime Human Factor Research Unit at Chalmers University of

Technology. Her main responsibility is research within work environment, human centred design and workplace design in the light of the technological changes and development that the shipping industry has and will face. She is a former Marine Engineer, served in the Swedish Merchant Navy and holds a PhD in Shipping Technology from Chalmers.

Scott N. MacKinnon is a Professor of Maritime Human Factors at Chalmers University of Technology. His current research is related to human-machine interactions in complex socio-technical systems. Previous to coming to Sweden, he was a Professor at Memorial University of Newfoundland where he conducted research on the impact of simulation and virtual reality upon knowledge and skill acquisition in maritime and offshore oil and gas industries. He holds a PhD in Biomedical Engineering from the University of Cape Town.

PRESENTATION

CONTOURS OF A ROADMAP FOR DEVELOPING SCENARIOS FOR THE WORLD OF AUTONOMOUS OCEAN SYSTEMS

S.Komandur and **T.Porathe,** Norwegian University of Science & Technology, Norway.

SUMMARY

There are three approaches currently with regard to predicting developments and anticipating changes in the world of autonomous ocean systems.

1. Prediction by a team of experts through discussion and consensus.
2. Building a prototype (Yara-Birkeland).
3. Simulators as the metaphorical crystal ball to gaze into the future.

A common thread amongst all these three approaches is the need for a background story (or) a *scenario*. It is safe to say that such a scenario must be realistic to be of any worth in predicting the transition from our current ocean system to this new world of autonomous ocean systems. *Delphi* method has long been used as a systematic approach to identify such scenarios. In this paper we suggest a modification to the *Delphi* method (*H-Delphi*) which we hypothesize is more suitable to develop scenarios for the world of autonomous ocean systems.

1. INTRODUCTION

In transportation studies the *Delphi method* has been widely used for predicting future scenarios [1]. Some studies have shown that a narrative approach to scenarios in development of socio-technical systems lends greater authenticity to such scenarios [2]. Some studies particularly from a business development context (not a transportation context) detail very specific approaches to scenario development, planning and alternate futures [3]. The *Delphi* method appears to be also useful in predicting unforeseen future scenarios in the logistics industry [4]. A detailed study also classifies the methodologies for scenario planning (which is different from scenario development) into a finite number of schools of techniques [5]. Advanced statistical methods such as cluster analysis are suggested to ameliorate widely criticized issues in the traditional *Delphi* method resulting in a new method called the disaggregated *Delphi* method [6]. A key ingredient of the *Delphi* method is identifying experts; one study proposes rigorous guidelines for selecting appropriate experts [7]. Given the long history of the *Delphi* method a summary of the lessons on the use of the techniques and some suggested improvements are documented in a paper [8]. There are many other known approaches to developing scenarios, one such example is *Shang enquiry* which was proposed as an alternative to the *Delphi* method in the 1970s citing many advantages [9]. Despite this the *Delphi* method has survived major criticisms as documented by a study [10]. It is important to note that *Delphi* method itself gained prominence by taking a critical position about the standard approach which is a consensus arrived at by discussions between experts in a particular domain [11]. Thus while noting the popularity of the *Delphi* method particularly in its approach to collecting and synthesizing experts' opinions, one must acknowledge that the 'round-table discussion format' (traditional format) despite its huge downsides [11] does have the advantage of using a lot less time for completing the process (i.e. arriving at a consensus). In a fast changing world of technology, elements of robustness from the *Delphi* method and the alacrity of the traditional method are both necessary for envisioning and manifesting the contours of the emerging socio-technical system. This is all the more true for ocean systems with the advent of autonomous entities within them.

2. METHODS

We propose a hybrid approach that combines characteristics of the traditional method and the *Delphi* method. Our hypothesis is that this method will save considerable amount of time with regard to the process and yet retain advantages of the *Delphi* method as compared to the traditional method, which made it popular compared to the traditional method in the first place. In both the traditional method and the *Delphi* method the following three steps are present; they differ mainly in terms of how they treat the third step.

> *1) Selecting experts wisely*
> *2) Providing proper conditions for their optimal performance*
> 3) *If there are several experts for an issue/domain one must use caution before deriving a single opinion from their various opinions.*

We propose following the exact same approach suggested by Helmer in his paper for the first two steps as described above [11]. For the third step we suggest the *Delphi* method as described by Helmer [11] instead of the traditional method albeit with a modification as described below. The modification suggested is to include a preliminary form of committee activity within small groups instead of the entire group of experts. For e.g. if there are 20 experts then the modification suggested is that 10 groups with two members each will be randomly formed and there will be committee activity within each group about the projected future and the consensus reached within each group will be the starting point of opinions instead of the opinions of each & every individual expert. From this point onwards the process is exactly the same as the Delphi method as described by Helmer [11]. For simplicity we are naming this approach *H-Delphi* (Hybrid-Delphi) method. A group of only two members for the discussion part is proposed due the following two reasons. First, we do not have a very large number of experts on specific domains of importance for ocean transport systems unlike many other domains. Secondly, smaller groups (i.e. n=2, 4, 6) tend to have fewer differences in the outcomes of discussions whether they use electronic or non-electronic approach for verbal brainstorming [12]. Thus it is possible to pair experts who are not geographically co-located. This also increases the possibilities for acquiring a large sample size during an empirical study comparing these methods.

If the visions emerging from the three methods (traditional, *Delphi* and *H-Delphi*) are relatively similar, then in itself it is a metric of the confidence we can place on the vision of the future state of the ocean system evolving from these approaches. On the other hand if the visions emerging are vastly different then it is important information for decision makers to reflect and subject all the three visions to greater scrutiny before choosing anyone of them as the guiding framework for the contours of the new ocean system with autonomous entities. This is particularly important due to the enormous value of the investments in this field.

3. EXPECTED RESULTS

We expect the visions/projections of the future emerging from the *H-Delphi* method will be a nuanced version of the visions emerging from the *Delphi* method. We expect the visions will only agree with those from the traditional method only when the visions are similar to those from the *Delphi* method itself. We expect the time consumed during the process using the *H-Delphi* method will be considerably lesser than the time consumed during the process using the *Delphi* method.

4. CONCLUSIONS

This is a work in progress paper and hence the conclusions must wait for a comprehensive empirical study comparing the three methods.

5. ACKNOWLEDGEMENTS

Our inspiration for this work in progress paper is the HUMANE project funded by the Norwegian Research Council. The authors thank the project leader and leading Maritime Human Factors specialist Professor Margareta Lutzhoft for her unstinted support.

6. REFERENCES

1. Melander, L., 'Scenario development in transport studies: Methodological considerations and reflections on Delphi studies', *Futures,* 2018.
2. Rasmussen L.B. (2008) 'The narrative aspect of scenario building - How story telling may give people a memory of the future', *In: Gill S. (eds) Cognition, Communication and Interaction. Human-Computer Interaction Series,* 2008.
3. Bishop P., Hines A., Collins T., 'The current state of scenario development: an overview of techniques', *foresight*, 2007.
4. Heiko, A., & Darkow, I. L. 'Scenarios for the logistics services industry: a Delphi-based analysis for 2025'. *International Journal of Production Economics, 127(1), 46-59.*2010.
5. Bradfield R, Wright G, Burt G, Cairns G, Van Der Heijden K. 'The origins and evolution of scenario techniques in long range business planning', *Futures*, 2005.
6. Tapio P. 'Disaggregative policy Delphi: Using cluster analysis as a tool for systematic scenario formation'. *Technological Forecasting and Social Change*, 2003.
7. Okoli C, Pawlowski SD. 'The Delphi method as a research tool: an example, design considerations and applications'. *Information & management*, 2004.
8. Rowe G, Wright G. 'The Delphi technique: Past, present, and future prospects—Introduction to the special issue', *Technological Forecasting and Social Change*, 2011.
9. Ford DA. 'Shang Inquiry as an alternative to Delphi: Some experimental findings', *Technological forecasting and social change*, 1975.
10. Gupta UG, Clarke RE. 'Theory and applications of the Delphi technique: A bibliography (1975–1994)', *Technological forecasting and social change*. 1996.
11. Helmer O. 'Analysis of the future: The Delphi method', *Technical Report RAND Corporation SANTA MONICA CA,* 1967.
12. Gallupe RB, Dennis AR, Cooper WH, Valacich JS, Bastianutti LM, Nunamaker Jr JF, 'Electronic brainstorming and group size', *Academy of Management Journal*, 1992.

7. AUTHORS BIOGRAPHY

Sashidharan Komandur is currently Associate Professor in the Department of Design at the Norwegian University of Science & Technology. He currently leads the Safety Critical & Autonomous Systems research in the Gjøvik campus of NTNU. His particular focus within this area is Maritime human factors. He was the Group leader of the Maritime human factors laboratory at the

Ålesund campus of NTNU which he started and managed from 2010 to 2015. He has been a key member of a number of research projects funded by the Norwegian research council and National Institute of Occupational Health (USA).

Thomas Porathe currently holds the position as professor of interaction design at the Norwegian University of Science and Technology in Trondheim. He has for many years worked with design of information systems within e-Navigation and Human Factors and related projects. He is active in the IALA e-Navigation committee.

PRESENTATION

SURVIVABILITY, ESCAPE AND EVACUATION: DESIGN FROM A HUMAN USER PERSPECTIVE

D Barber, UK

SUMMARY

The paper examines systems in common current use aboard merchant ships, their regulation and practical aspects of their deployment. The issues examined include not only life saving devices but the platforms from which they are deployed and the likely scenarios in which they are expected to function. The examination is conducted from the viewpoint of a mariner with input from experience in practice. Types of vessels are separated to reflect the differences in behaviour that should be expected, based on scientific analysis as well as practical experience of meteorological and hydrological influences. Human reaction is reflected within the descriptions. (maximum 150 words).

NOMENCLATURE

BIMCO Baltic International Maritime Council
CHIRP Confidential Hazardous Incident Reporting Programme
Davit A type of lifting appliance used (in this paper) for lowering and recovering boats to and from the water.
EFSWR Extra Flexible Steel Wire Rope
Fall Rope, usually wire, that supports a boat beneath a davit.
FPD Fall Preventer Device
HMPE High Modulus Polyethylene
ILG Industry Lifeboat Group
IMO International Maritime Organization
LSA Life Saving Appliances
MES Marine Evacuation System
NGO Non-Governmental Organisation
RINA Royal Institution of Naval Architects
SOLAS Safety of Life at Sea Convention
Survival Craft Any craft such as a lifeboat, liferaft or other vessel capable of providing a buoyant platform for persons evacuating a parent vessel.
VLCC Very Large Crude Carrier
Capesize (Bulk Carrier) Bulk Carrier with a beam greater than 32.3m (106ft) (Panama Locks pre-2016) Typically the range of deadweight tonnage for Capesize is from 90,000 tonnes to 200,000 tonnes

1. INTRODUCTION

This paper has its origins in the experiences of the author, both serving at sea and in management ashore and as an inspector of ships for both commercial principals and flag state. The author can recall his first lifeboat stations exercise and thinking at the time "this could be dangerous". After 52 years, he is disappointed that his view has not changed. Indeed it has become even more pronounced as lifeboat designs have been increasingly mechanised and the number of casualties during training exercises has escalated. Frustration at the slow response of regulators and the lack of an official central database of accidents and near misses have only served to strengthen his resolve to try and influence change.

RINA has recently joined the Industry Lifeboat Group (ILG) – a group of interested parties consisting of ship owners and operators from around the world, seafarer representation with similar coverage, Insurers' representation at an international level and now design representation. RINA therefore has an opportunity to join the chorus for change where it is needed, adding its unique and much needed credentials to the mix of influences around the ILG meeting room..

SOLAS tends to assume that the current approaches to evacuation are fit for purpose. This paper questions whether such an assumption is safe. A number of key factors in vessel behaviour and practical approaches for the participants in the evacuation have been neglected, misunderstood or at worst, denied.

2. ISSUES TO BE EXPLORED

2.1 EVACUATION AND ESCAPE

For the purposes of this paper:

- Evacuation means removal of personnel from the vessel, not from its constituent parts such as from a compartment or compartments within the ship;
- Escape means removal of survival craft and/or personnel in the water from the immediate area of the vessel to a position unaffected by the hazards directly associated with the stricken vessel; and
- Survival means the continued sustaining of life amongst the personnel affected by the evacuation.

2.1(a) Methods in Use

It is commonly understood in the industry, from designers through to end users that the principal methods of evacuation are as follows:

- Lifeboats, of which there are various types launched (and recovered) by lifting devices known as davits incorporating usually wire rope falls. Of these there is a further subsection known as Free-fall lifeboats, which are used extensively in the offshore industry but also aboard tankers and an increasing number of cargo ships;

- Liferafts, of which there are multiple forms but most conform to a fairly standard design of round, oval or rectangular, usually inflatable lower sections with a canopy above. There have been rigid designs but these are not in widespread use. The rafts are designed to be thrown overboard or to float free and automatically inflate in the water. Some can be lowered from davits after inflation and embarkation at deck level;

- Lifejackets, either permanently buoyant or inflatable and conforming to internationally agreed standards that ensure the wearer is sustained floating in a face up attitude once in the water;

- Life Rings, - also known as lifebuoys - consisting of a buoyant ring that can be thrown to persons in the water. Like other gear they are subject to internationally agreed standards;

- Other, innovative methods, which may be devised at the time by users (when there is difficulty in deployment as intended by design) or alternatives approved by administrations under the provisions of SOLAS Chapter I Regulation 5 that provide floatation and ideally shelter for the users.

SOLAS wording prioritises lifeboats as the principal means of evacuation. The impression created is that whilst lifeboats are trusted, innovations are not, despite the existence of Regulation 5.

2.1(b) The Parent Vessel

The paper assumes that the parent vessel is a trading vessel in international waters. Trading vessels fall into the following major types:

Cargo Vessels

With the exception of very small coasting vessels, most cargo vessels carry the same types, albeit varying sizes of lifeboats. The principal difficulties with lifeboats are not so much with the boats themselves however but when trying to launch and recover them. The distance to the water can introduce very real problems as freeboard increases not help by a widespread distrust of the connections to falls after many years of continuing accidents.

An increasing number of vessels carry the so-called free-fall lifeboats that were once only likely to be seen aboard tankers. Free-fall boats do introduce different challenges and although not noted for fatal accidents have experienced a number of recorded near misses, which should be ringing alarm bells. The argument in favour of these boats because of the low number of fatalities ignores the fact that they are not required to be exercised to the same extent as conventional boats and when they are, because of the acknowledged hazards present, they are usually under more intense supervision..

Passenger Vessels

A significant proportion of passenger vessels now carry many thousands on board. Seafarers can be fearful of having to evacuate such large numbers. A number of masters and others have expressed concerns in professional fora. This is not because the crews are untrained to deal with the passengers – they now undergo training for crowd dynamics – but because they are fearful of the operation first of getting the lifeboats and other LSA loaded with untrained and frightened people and then, as with cargo vessels, successfully launching them without mishap. They are aware that exercises are only ever carried out in the most benign conditions in port, with limited numbers of fit, trained seafarers in the boats and with the ship upright and level keel. In other words they are aware that the systems that have evolved since the TITANIC are, in effect, unproven in real open sea conditions. Casualties that have occurred have often had major difficulties with launching both lifeboats and liferafts and once in the water the problems compound as overcrowding turns quickly from an issue of comfort to one of survival. IMO is currently discussiing how to distinguish between comfort and survival. The discussion, which concerns itself with special metrics, ventilation and sanitation (though not much with the latter) is limited to the boats and concerns itself less with launching gear. As with any committee discussion, the "terms of reference" of the debate are deliberately limited, a luxury not afforded to a master and his/her crew in an emergency.

2.1(b) Approaches of Legislation

Since inception post-TITANIC, legislation has become increasingly prescriptive in what is required of life saving appliances in a style that has a tendency to apply common requirements across widely varying vessel types. In this "one size fits all" approach unintended consequences are more likely to occur. Manufacturers understandably seek a common standard to which they can mass-produce equipment. Pike, [1] alludes to a "race to the bottom" in design, particularly in relation to davits as a result of prescription.

Assumptions are made concerning the attitude of the vessel without questioning whether these assumptions represent a truly realistic scenario. The assumptions go unchallenged because most equipment is only ever tested in calm conditions from upright vessels usually in port.

Regulation also tends to omit recovery operations. They are not considered part of any abandonment process. But SOLAS does concern itself with drills involving full launching and these drills must involve returning boats to their stowed positions. Drills occur at a much higher frequency than the design one way abandonment trip. There will be a heightened risk due to the increased frequency of exposure. This is possibly a significant factor in why the fatalities known about so far have occurred during drills or maintenance.

IMO debates frequently raise the issue that drills should, be as realistic as possible and SOLAS itself requires it. True reality will most likely be accurately assessed by serving seafarers. A person ashore analysing the circumstances from a desk top or committee room is more likely to miss something. Inspectors and surveyors representing administrations, could, instead of assuming a position of superiority provide a valuable conduit for feedback from the seafarers themselves. How much are they willing however to challenge the effectiveness of current legislation in the deployment of life saving systems?.

2.2 TRIGGERS TO EVACUATION

2.2(a) Vessel Ashore.

Many seafarers, masters in particular, will state that the best lifeboat is the ship. This indicates a reluctance to leave it. This reluctance is often misunderstood, particularly by coastguards who have as their primary aim the safe evacuation to shore of all victims of shipwreck. The problem with this approach, though laudable, is that it is rooted in the 19th century when ships were made of wood and would disintegrate as soon as they became stranded. Modern all-welded steel hulls tend to hold together rather better and many has been the occasion – not always - that on the morning after the storm the steel vessel is ashore, possibly broken into sections, but with refuges such as an accommodation block above water, in one piece and still habitable. Against this the crew might still be lost, having possibly mistakenly believed through the way they have been trained, that their best option was to leave the vessel, when in fact such action would lead them to a much more dangerous place.

2.2(b) Vessel Sinking – Foundering.

If the stricken vessel has limited time in which it is able to stay afloat and/or upright, clearly there is a need to evacuate before it ceases to exist on the surface. The problem is "when and how" decision that must be made by the master. Most masters will be reluctant to commit any crew or, even more so, passengers to survival craft unless they are sure that the vessel cannot survive, which is often difficult to determine. An additional cause for delay can be the natural reluctance of masters to commit to abandonment through distrust of the equipment. There is a very real danger therefore that the wait may lead to a situation where evacuation is less feasible. It is wrong to ascribe the delay to indecision. There are likely to be multiple other decisions being made at such a time and it may simply be a case of responsibility and information overload.

Mechanism of Sinking

The mechanism of a vessel sinking will be governed by the integrity of the hull sections to be flooded, their size relative to the overall displacement of the vessel and their disposition.

Cargo Vessel Sinking

Most modern cargo vessels have engine rooms aft. Large container ships are now tending to have engine rooms towards but still abaft amidships.

If the engine room is flooded, although the result will be very expensive and probably totally disabling, it will not usually sink the vessel without breaching one or more additional compartments.

If however the cargo areas are breached a situation can be rapidly reached, through progressive flooding, that cannot be survived. The loss of buoyancy will, in most vessels be well forward and the trim will immediately be by the head to the extent that vessels rapidly trend towards complete upending – head down, stern up (**Figure 1**). Breaching of the after cargo compartments of vessels configured like large containerships may well lead to a trim reversal (i.e. by the stern) but the volume available for flooding may, like the engine room, be insufficient to take the situation to one of progressive flooding because of the finer lines of the hull form in that area. Larger vessels however are likely to suffer structural failure as the loss of upthrust places increasing loads on the hull girder.

Figure 1: Upending during sinking. ERICA (Tanker) Sinking. Note conventional lifeboat floating free
Source Internet.

Passenger Vessel Sinking

Unlike cargo vessels, passenger vessels have much more complex subdivision in their hulls and their attitude in sinking can, to a limited extent, be controlled. Upending during sinking (**Figure 2**), whilst it does occur, is not a given.

Figure 2: Upending during sinking. OCEANOS Sinking off South Africa. Source Internet VideoSource.

The masters of passenger vessels have all the same reasons for reluctance to commit to abandonment as their cargo vessel colleagues but have the added pressure of dealing with often very large numbers of untrained participants.
SOLAS assumes an adverse heel as the worst barrier to evacuation but unlike cargo vessels passenger vessels do not have to carry enough LSA on one side of the vessel to accommodate the entire population on board. They therefore cannot simply move to the low side of the vessel to evacuate as is possible with a conventionally equipped cargo vessel. The master will be under pressure to evacuate early, which will be in direct opposition to his likely instinct that the best place could be on board the vessel and not in an exposed lifeboat, tossed about in a seaway with immense overcrowding. If it turns out that the vessel is truly doomed of course, and reader please spare a thought for the time (s)he has to make such an assessment, there is no choice.

2.2(c) Vessel on Fire or in Gas Cloud

Regulators appear to have a preoccupation with fire, which is not to dismiss its importance, but many large modern vessels can provide a safe haven on board, completely removing any need for abandonment. The fire scenario or the gas cloud, which is probably as likely, is used to justify the fitting of free-fall lifeboats to tankers, gas carriers and chemical carriers, the assumption being that such craft provide a fast and efficient means of escape. This assumption should be challenged.

Close examination can show that the lessons of the BRITISH TRENT disaster appear not to have been learned [2]. The nine lives lost were largely lost on deck, enveloped by smoke and heat from the fire. They could not reach the lifeboat. Many modern vessels can be seen with free-fall lifeboats perched atop launching ramps, the only access to them being across open decks and sometimes up several ladders. Tankers, Gas and Chemical Carriers have stringent design rules for keeping the accommodation gas fee. The rules however do not appear to apply to the location and access to the lifeboat(s).

2.3 COMPLICATIONS AND BARRIERS TO EVACUATION

Davit Design Evolution.
Trim is a limiting factor that may easily be overlooked. Modern davits have the two falls arranged such that they are wound from individual but linked drums. The davit head blocks of older designs have, on modern davits evolved into sheaves with fixed axles through the davit arm itself. The limitation on trim is thus very necessary to prevent displacement of the fall from the head sheave groove. Tilting of the boat matching the attitude of the vessel could result in serious injury or worse of the occupants. In earlier circumstances passenger vessels such as TITANIC (and indeed cargo vessels) had much smaller boats that could be launched from davits with fibre rope falls that could be individually controlled over a ramshorn at each davit (**Figure 3**).

Figure 3:. TITANIC Sinking. Source Internet. Scene from Film.

The attitude of the boat could thus be compensated to counter the trim of the vessel. Although labour intensive the older designs probably ensured a more orderly evacuation. On the newer, larger, more mechanised davits, this compensation for trim is no longer possible.

For low freeboard vessels such as cargo vessels up to about 20,000 tonnes deadweight the limited distance to the water is far less daunting than on a larger vessel such as a capesize bulk carrier, large container vessel, VLCC, vehicle or livestock carrier or some cruise vessels. Designs have tended to place the lifeboats and any other life saving appliances (LSA) on top of these superstructures. The prospect of somehow transiting the precipice can often involve a drop to the water of 20 metres (on level keel).

2.3(a) Lifeboats

The traditional basic survival craft is still the lifeboat. The principle has been based on the same logic since first formally adopted around the end of the 19th and beginning of the 20th centuries. The TITANIC disaster, which notoriously had insufficient numbers of lifeboats brought their use into focus and since that time the basic design has altered little except in its size, materials of construction, degree of shelter provided for the occupants and most recently and infamously, the means by which the craft are attached and released from the falls supporting them.

Lifeboats, being powered are able to provide means of escape for survivors from the immediate vicinity of the parent vessel.

.

Conventional Davit Launched Lifeboats

From their inception to the end of the 20th century the established means of launching lifeboats from parent vessels has been the davit. Essentially the davit is a type of crane. The usual arrangement is a fall, usually constructed of wire rope attaching the boat at its forward and after extremities to the davit(s) of which there are usually two – one at each end of the boat. Some smaller craft and related rescue boats use a single fall and davit or crane but for larger boats this is normally considered impractical. The two falls usually also serve a "luffing" purpose as topping lifts at the very top of the hoisting section to enable the davits to swing out – to luff – from the inboard stowed position to an outboard position from which they clear the hull (and superstructure) when lowered. This will remain possible only if the vessel remains upright or, if heeling, on the low side of the heel. On the high side the boats are expected to slide down the ship's side aided by the provision of "skates" attached to the boat hulls in such a way as to be jettisoned once away from the parent vessel.

Pendulous Oscillation

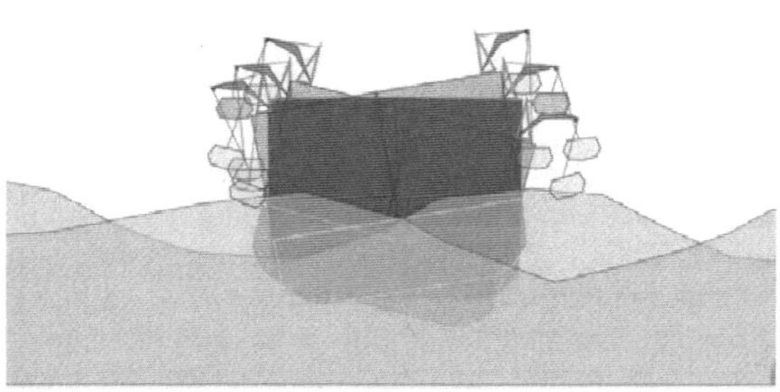

Figure 4: Pendulous Oscillation.

Boats descending clear of the parent vessel become part of a pendulum created by the extending falls. As the pendulum increases in length the boat's tendency to swing pendulously will increase, excited by even the smallest lateral movement at the pivot point – the davit head (**Figure 4**). As the pendulum length increases the period will approach synchronisation with the ships own rolling period. Such oscillation can have alarming effects for the occupants of the boat such as striking the parent vessel's hull or capsizing as the keel of the boat contacts the water at an extreme lateral velocity.

It is noted that whilst SOLAS considers heel and trim there is no consideration of roll pitch or heave and yet every vessel will immediately begin all of these motions as soon as its way through the water is slowed or stopped. The master will have no means of controlling the motions if the vessel is without propulsion power – the most likely scenario.

One option could be to reduce the pendulum length by using innovative design on the lowering gear such as is used ashore on knuckle cranes or excavator arms. Such devices could reach below deck level and taking advantage of modern hydraulic capabilities may even be able to eliminate the use of a fall wire completely, replacing it with a sling from the device end. Some small vessels have already achieved this with knuckle cranes on their smaller (usually rescue) boats.

Free-fall Lifeboats

Increasingly, as well as for vessels carrying dangerous cargoes, free-fall lifeboats have been fitted to dry cargo vessels and following the findings of the International FSA of Bulk Carriers they are now a requirement on those vessels [5]. In this latter case the recommendation also included a float-free capability to combat the rapid sinking that is known to occur with bulk carriers carrying dense cargoes if they suffer loss of hull integrity

Has the concept of free-fall boats been properly thought through? They are invariably launched over the stern. The regulations require that they can do so through all the same attitude variations of trim and heel as conventional boats – 10 degrees of trim, 20 degrees of heel in either direction - and yet has any such boat ever been fully tested at these attitudes and if so, how? Add to this the roll, pitch and heave mentioned above and there are significant interference factors that could affect the delicate balance that establishes the trajectory of the lifeboat. Can a free-fall boat continue to be launched as the ramp inclination is decreased by the increasing trim by the head during upending?

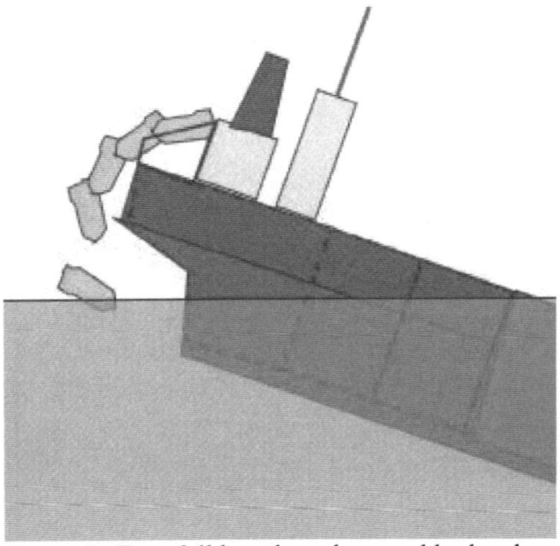

Figure 5: Free-fall boat launch, vessel by head.

It should be obvious that this trim change will quickly reduce the angle of slope on the launching ramp, which will in turn slow the launch velocity. Karim and co-authors in their research [3] showed that with diminishing speed of ejection the boat is likely to over-rotate (**Figure 5**) and indeed this possibility has been proven by events by an anonymous but actual event recorded on Youtube. It can be viewed under title front-flip lifeboat [4]. The event occurred on an upright vessel in port. The incident implicates known causes of impairment of the ramp velocity of the boat, which can be caused by deficiencies in roller freedom.

Float Free

Free-fall boats can float free – a feature recognised in the FSA for Bulk Carriers in IMO. They can be launched "up" the ramp by buoyancy as well as down it by gravity. The process need be very little different other than the attitude of boat experienced by the crew. Training for crews however does not recognise this fact and yet it is probably the most likely scenario. The delay possible, indeed necessary, would enable the master to utilise maximum time for evacuation, only finally departing the vessel as it departs from the surface of the sea. (**Figure 6**)

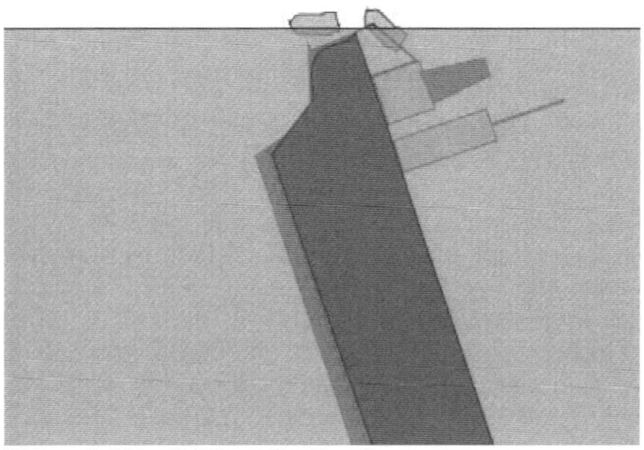

Figure 6: Free-fall boat floating free.

Questioning of crews by the author during inspection and drills has revealed that most are unaware of this float-free capability. More worryingly, many indicate that in such circumstances they would probably jump into the sea, believing the boat to be doomed along with the parent vessel.

What is there to prevent conventionally launched lifeboats to be held in cradles alongside or below boarding decks? They could float free when the vessel finally sank; the time that it ceases to be the best lifeboat.

2.3(b) Liferafts

Evolved from Carley Floats of WWII, liferafts have been the most dominant alternative survival craft aboard ships since that time. SOLAS however still regards them as secondary. More recent evolution has made these craft a very real substitute for lifeboats for many vessels.

Liferafts are not powered but can be towed by lifeboats or marshalling craft to escape the immediate vicinity of the parent vessel.

There are two types of liferaft:

Throw-overboard

These rafts remain packed in their cases until launched to the water, where they inflate. They cannot be boarded until inflated therefore, by definition, they have been designed with boarding from the water in mind.

Personal Survival Course participants will know just how difficult the raft is to board even in the highly controlled environment of the training pool. Improvements in design, include a small immersed step or boarding ramp by at least one entrance to facilitate more effective boarding. The step however does not overcome the high risk of hypothermia during the time in the water when survivors will grow quickly weakened.

Davit Launched

Dry-shod evacuation from the parent vessel would require boarding the liferaft before it enters the water. This is possible if the raft is davit launched or in very low freeboard vessels by direct boarding. The liferaft will incorporate reinforcing of the raft and canopy to enable suspension from a davit fall. Ferries and passenger vessels are frequently equipped with these types of liferaft as are fewer cargo vessels. They were successfully used in the ESTONIA rescue for recovering persons from the sea. [6]

During launching the pendulum effect will still exist as it does for lifeboats, especially since davit launched liferafts are usually stowed on decks well above the waterline but the craft are better able to absorb impacts than are lifeboats and except in very high wind speeds, are more stable on the water surface.

Boarding Issues

Dry boarding is not a method of boarding envisaged for "throw overboard" type rafts, which number the majority in cargo vessels and many inshore passenger vessels. The exception is when such rafts are the survival craft associated with a Marine Evacuation System (see below).

There have been demonstrations of inspired inventiveness by some crews. For example, when the TRICOLOR – a vehicle carrier involved in a collision in Dover straight – was rapidly capsizing and sinking, the crew did not waste time trying to launch lifeboats. Instead they inflated a liferaft on the top deck and climbed into it. The raft floated free as the vessel capsized and there were no fatalities.

Regrettably there appears to be no official recognition of this method of boarding.

What is there to prevent all liferafts being inflated at deck level, possibly on platforms that emerge from the side? They could then be boarded dry shod and would float free when the vessel finally sank; the time when it ceases to be the best lifeboat.

2.3(c) Marine Evacuation Systems

Marine Evacuation Systems (MESs) differ from all other systems envisaged by SOLAS in that they must be tested in a seaway in conditions equating to wind speeds of force 6 and significant wave heights of at least 3 metres – a realistic test.

MESs are usually inflatable and are therefore unsuitable for use where fire on the water could be an issue.

There are two fundamental types of MES:

Slide types

Figure 7: Slide MES during Sea test.: Source SURVITEC Internet Publicity Film.

Found mainly on ferries, slide type MESs have also been used on a few, usually smaller, cruise vessels. Having been used in aviation for many years the system is well proven and developed.

The simplicity of the designs is a strong attribute. Persons gripped by fear, disabled and stretcher cases can be lowered down the slide on lines and accompanied by crew members and families can remain together, which strengthens morale.

The lower end of the slide is visible throughout, which avoids collisions and enhances confidence. Some versions have reinforcing tubes that form a partial enclosure above the slide (**Figure 7**). This creates a greater sense of separation from the sea and reduces the tendency towards visually stimulated vertigo.

Most users will be familiar with the concept from infancy (Figure 8).

Figure 8: MES Trainer. Photo D. Barber

The slide also has one strong feature in that it can be and has been used in the opposite direction for rescue of persons in the water as dramatically demonstrated by the master of the ISABELLA in the ESTONIA rescue [6].

It is encouraging to note that IMO has recognised this method of rescue [7]

Chute types

A second form of MES has developed, particularly on large passenger carrying vessels, including some ferries. This involves a vertical chute between the deck of the parent vessel and the boarding point of the liferaft. Early designs incorporated a series of elasticated apertures or linings (similar to a *tubigrip* bandage), through which users are expected to jump. The fall of the user is arrested by the tension of the aperture on their body. Presumably persons of greater mass pass through more quickly, opposed only by a marginal increase in friction due to increased surface area in contact!

There has been one fatality using this type of chute in which a participant became trapped in a "piked" position, i.e. doubled with feet and head together, caused or aggravated by the lifejacket type worn. The rescue failed to save the participant before death occurred due to suffocation. The accident investigation led to the introduction of "sweepers" - trained members of crew who will enter the chute to clear any such blockage.

The bottom of the chute is not visible from the top. It is suggested this may act as a fear factor in untrained persons, not to mention difficulties in evacuating disabled or injured persons by such a system.

Another descent control used in chute types is the incorporation of a spiral slide inside the chute. Whilst being similar to the open slide the bottom is still not visible from the top so untrained and possibly very nervous passengers cannot be expected to use this type of slide with enthusiasm.

On the internet it is possible to find other imaginative versions of the vertical chute including at least one that looks as though intended for offshore platforms in which participants bounce from alternating trampoline like surfaces (and in some cases become entangled). It is difficult to imagine an elderly person, particularly with potentially brittle bones negotiating such a descent without sustaining injury and one should not even consider disabled or stretcher borne casualties.

2.4 MAINTENANCE ISSUES

2.4 (a) Wire Ropes

Lifeboats have traditionally been supported on wire rope falls ever since the boat size grew to a point that the previous practice of fibre rope falls necessitated the additional strength inherent in steel wire. The rigid rule on wire rope use during the author's early experience was that running rigging (e.g. falls) should always use what was then described as extra flexible steel wire rope (EFSWR). EFSWR included a fibre core that gave it greater ability to deform when the wire rope was led around sheaves and rollers. The deformation would compress the oil soaked fibre core thereby delivering lubricant to the wires surrounding it.

Less flexible, steel core wire rope was, under the same rule, strictly limited to "standing rigging" applications such as stays and shrouds, As well as being less flexible it was also unable to automatically re-lubricate from the fibre core it lacked. Stays and shrouds would have been coated with a harder coating aimed at keeping corrosive influences out.

With the development of very large lifeboats and tenders on passenger vessels the use of steel core wire ropes in running rigging (falls) appears to have become normalised. It is claimed that frictional wear between the strands is alleviated by "pre-lubrication" – a factory process involving forced lubricant into the ropes by a pressure jet process. Such a specialised process is not normally possible onboard ship. A number of recent accidents have implicated corroded and failed wire ropes. Lack of lubrication is regularly cited in reports of such accidents, the protective galvanisation having been abbraded away exposing the steel to the aggressive environment at sea.

Fatigue should also be considered. Each time the wire rope is bent, such as around sheaves, of which large boat davits have many, it is subjected to stress reversals that are the principal cause of fatigue. As the number of reversals increases so does the fatigue. Typically, when a wire rope is removed from its davit installation, it is invariably deformed, confirming that it has exceeded elastic and entered plastic deformation, i.e. it is fatigued and reduced in strength.

Alternative Materials

Other materials such as HMPE have emerged as a viable alternative for falls and have been used extensively in the offshore industry for other purposes as well as on cranes ashore, but some incidents under investigation have created a mistrust of these materials. The extensive and convoluted approval process makes it difficult to introduce alternatives and the status quo is likely to continue unchallenged, regardless of whether it is implicated in accidents.

Suitability of Deployment

Wire rope is stiff and lacks elasticity, which should limit its range of deployments. Launching gear on lifeboats involves many functions that frequently induce shock loads. Safety strops, designed to arrest any movement in the event of premature or inadvertent release of securing arrangements, have often been found constructed in wire rope, or worse still chain. It would be relatively simple to replace wire rope in these functions and with lifting bridles using resilient fibre strops or slings, as is done for Fall Preventer Devices (FPDs) – intended to provide an alternative load path when on-load release fails. A number of accidents involving failure and one in particular [9] that highlight this deficiency.

Resilient fibre recovery strops or slings between boats and falls would also alleviate shock loads associated with the pick up from the water. One reason for reluctance to launch boats might be the shock loads that occur on the release hooks, falls and davits.

Davits and Launching Ramps

This equipment incorporates numerous sheaves or rollers, all of which have bearings that require maintenance and regular proving. Many designs – some would say most – do not pay much attention to the servicing function that should be possible on board ship and the designs frequently do not even incorporate accessible greasing points.

3.0 CONCLUSIONS

The observations included above are not intended to be a diatribe of complaints. They are intended to be an alert to designers that existing arrangements for life saving aboard ship could be improved.

The human reaction to existing arrangements usually translates into a reluctance to use such equipment for exercise purposes and it is intended that the descriptions herewith should serve to explain the reasons for such reluctance. Further to this however is the additional stress on those in authority who, having extensive experience, can identify many of the same issues as are highlighted in the paper but feel powerless to become involved in any improvement. The stress is created by the responsibility that all such persons have in their duty of care to their crews and passengers.

The many references to SOLAS could create the impression that it is considered unfit for purpose, which is far from the actual view held. It is however in need of major revision but that may take several years. In the intervening time it is hoped that the provision allowed within SOLAS for innovation and alternative design could be more extensively used, which logically should also necessitate simplification of the formalities necessary to obtain exemptions.

Practicing professionals should be reminded that SOLAS represents a minimum standard. There is nothing to prevent that standard being exceeded other than commercial competition with cost implications – the "unlevel playing field" - or ignorance of reality.

Discourse between serving seafarers, operators, manufacturers and designers to identify where improvements can be made should be more rigorously encouraged. Some innovations might actually be cheaper than what exists at present.

Above all, the industry, from design to operation, should review its goals and compare those to the effectiveness of what is available, seeking to eliminate unnecessary risk and make familiarisation of equipment less onerous to the seafarer.

In a comprehensive study of lifeboats made in 2001, Admiral John Lang, the then Chief Inspector of Accidents at the UK MAIB recommended that IMO should undertake a study on the present value, need, and desirability of lifeboats. Reported incidents and accidents worldwide should be examined with regard to the specification of lifeboat launching systems and whether or not they are actually necessary.

Although there have been numerous measures introduced that have been aimed mainly at procedures, to date, the author is unaware of a comprehensive study addressing the specific point raised by Admiral Lang - the continued feasibility of lifeboats as an effective means of evacuation and escape.

4. ADDITIONAL INFORMATION

4.1 DISCLAIMER

The content of this paper is formed by the personal observations and opinions of the author. Nothing in the paper should be interpreted as the opinion or policy of any organisation with which the author has been employed or associated throughout his entire working life.

4.2 LIMITATIONS

The necessary limitations of the paper have inevitably excluded some detail that could further enhance the descriptions given. The author emphasises that he is open to further enquiry to elaborate on content of the paper or on any associated issue that may have been excluded.

5. REFERENCES

1. PIKE, D.'Launch and Recovery of Boats from Ships', *Nautical Institute,* 2018
2. UK MAIB, Report of the Investigation into the Collision between, BRITISH TRENT and WESTERN WINNER with the loss of nine lives. ,*MAIB 1993.*
3. KARIM, M.M et al, 'Numerical Investigation into the Effect of Launch Skid Angle on the Behaviour of Free=Fall Lifeboat in Regular Waves', *Trans RINA, Vol 151, Part B1, Intl J Small Craft Tech, 2009 Jan-Jun*
4. ANON, "Lifeboat Launch Front Flip win', *Internet Youtube*
5. IMO MSC, Survival Craft, *SOLAS Chapter III, Regulation 31.1.8.*
6. ESTONIA GOVERNMENT, 'Final Report into the Capsizing on 28 September 1994 in the Baltic Sea of the R0-Ro Passenger Vessel MV ESTONIA', *The Joint Accident Investigation Commission of Estonia, Finland and Sweden,* 1997.
7. IMO MSC, 'Means of Rescue', *SOLAS Chapter III.Regulation 26.4.2 and 3.*
8. UK MCA, 'Operational Issues relating to Marine Evacuation Systems', *MGN 273(M) 2004.*
9. AUSTRALIAN GOVERNMENT, 'Unintentional Release of the freefall lifeboat from bulk carrier *Aquarosa',Australian Transport Safety Bureau.*

6. AUTHORS BIOGRAPHY

Captain Dennis Barber is now nominally retired but remains Associate Partner of Marine and Risk Consultants (Marico Marine) in which capacity he is an Approved Nautical Inspector for the Bahamas Maritime Authority. In his previous professional life he was a seagoing officer with P&O Bulk Shipping as well as Offshore Marine and Sealink Ferries. He was a Cargo, Marine and finally Chief Marine Superintendent with P&O Bulk Shipping between 1980 and 2000. He was then contracted through Marico Marine to UK MCA as Project Manager for the DEBYSHIRE Recommendations and was part of the team managing the International Collaborative Formal Safety Assessment of Bulk Carriers at IMO between 2000 and 2004. He was a founder member of the Industry Lifeboat Group as the representative for BIMCO with whom he has been a technical adviser to its Marine Committee from 2003 to the present.

He is a member of the CHIRP Marine Advisory Board.

He is a Fellow of the Nautical Institute, Member of the Royal Institute of Navigation and an Associate of RINA.

He is a Freeman of the Honourable Company of Master Mariners and a Younger Brother of Trinity House.

PRESENTATIONS

A Virtual Teams Model for Supporting Maritime Technology Management

G Berner University of Tasmania, AUS
R Hopcraft Royal Holloway, University of London, UK
J Scanlan University of Tasmania, AUS
M Lutzhoft, Western Norway University of Applied Sciences, Norway
J Earthy Lloyd's Register, UK

SUMMARY

Computer and networking technology on board ships is increasing in complexity as the levels of automation and monitoring at sea evolve. Ships are acknowledged to be a System of systems, including both IT (Information Technology) and OT (Operational technology). A virtual team that includes members of the various on-board departments, shore based technical support staff and vendors perform management of technical equipment, largely in isolation. It is possible to conclude that the technology installation is one of the only operational aspects of a vessel, where no one has the full picture and could drive a coordinated response to a major technology issue. This paper will propose a best practice framework for governance of technology on board ships. This framework will serve as an input to the process of including cyber security in the ISM Safety Management Manual for the 2021 deadline.

1 INTRODUCTION

The increasing growth of Information Technology (IT) and Operational technology (OT) on board sea going vessels is a response to the environment of increasing competition and regulation in combination with a reduced complement of seafarers. This has resulted in a developing reliance on integrated technology systems for day-to-day ship board operations [11]. The use of technology also supports the expansion of shore based monitoring and control [15]. The use of shore based personnel to perform support functions for ship based technology is a growing trend in maritime technology management where the responsibility of operating and maintaining maritime systems resides with a "virtual team" including members of the vessels crew, shore based operators and technology vendor staff. Achieving the goal of reliable technology in any field of endeavour requires well-designed solutions that are maintained according to good engineering practice with appropriate governance systems overseeing their operation. In many cases the management of technology on board ships could currently be characterised as being ad-hoc. The technical capability can be represented as a collection of islands of expertise/practice with little standardisation, making them worthwhile candidates for maturity uplift through the implementation of IT Service Management (ITSM) [5]. IT Service Management has been identified by [5] as a method for providing stronger alignment between technology and its consumers, improved service delivery and utilization of resources.

Robust IT governance systems have demonstrated that it is possible to reduce the potential for operational failures in critical systems [7]. A failure of critical technology can endanger life, property and the environment. An example highlighted by [16] demonstrates a lack of cohesiveness between a virtual technology support team and consumers of the technology service placed a vessel at risk. In this case, a software patch was installed remotely by a service engineer who was unaware of the operational context of the vessel at that time. The vessel was the process of berthing and the install resulted in a system reboot that stopped the engine room ventilation system, and subsequently left the vessel without propulsion at a critical time while it was manoeuvring.

Since the 1990s technology organisations have been using governance frameworks such as the IT Infrastructure Library (ITIL) [7] to provide significant benefits to their respective organisations through rigorous control of testing and system changes, resulting in more predictable infrastructure, reduced faults and consistent handling of technology incidents. Organisations who have adopted these technology service management frameworks enjoy cost effective services, reduced downtime, improved security and higher customer satisfaction [5]. This paper will discuss the potential for increasing the maturity of technology service provision on sea going vessels through the use of an IT governance framework.

2 MARITIME TECHNOLOGY SYSTEMS MANAGEMENT AND ITS CHALLENGES

While there are currently many technology companies providing shipboard solutions to the maritime industry, historically the industry has lagged behind other shore based industries in adoption and governance of new technologies. The global focus on cyber security over the last few years has led to activity in the development of guidance and regulation. This includes the IMO releasing requirements for management of cyber risk [9], and various other maritime organisations releasing technical papers and guidelines - [3], LR [14], DNV [6]. There are also a number of standards being developed that address maritime technology, The International Standards Organisation is currently drafting ISO-19847 and ISO-19848, specifying minimum standards for on-board computing and communications. Despite these recent efforts, the overall current maturity of IT service provision is low, and time and commitment are required from stakeholders to reach a suitable level. The following sections discuss some of the challenges ahead.

Technology management on sea going vessels can be described as having a relatively low level of maturity. Two well-known aspects of this are software quality [2] and cyber security [9]. These challenges are symptoms of a unique and complex operating environment that often features different stakeholders managing parts of the technology infrastructure in isolation. Company technologists, seafarers and vendors all participate in supporting and operating the specialised equipment that is installed on-board. There are however few, if any, processes around over-arching technology governance.

A new vessel entering service is designed for an operational lifespan of 25 years [8]. Technology systems, both hardware and software are supported by their respective vendors for only short period of time relative to the lifespan of a ship. Once hardware and software reaches it's 'end of support' date, software and firmware upgrades are no longer provided reducing the reliability of the technology. Many vessels in the worldwide merchant fleet rely upon legacy hardware and software that is no longer supported by vendors, placing them at increased risk of technology failure.

With respect to configuration management the nature of the technology installation is such that changes to installed software, hardware or configuration are typically made directly into the 'production' system, as there is often no test network available that would allow for full validation of the change prior to implementation. This is a significant deviation from best practices used ashore when making changes to production [12].

2.1 COMMUNICATIONS

Vessels travel around the globe and are often reachable only via low bandwidth / high cost satellite communications, limiting the opportunity for remote shore based technical support. Due to the limited physical access to systems the downloading of updates and patches increasingly needs to be performed remotely via a limited bandwidth connection [14]. The size and release frequency of software patches for operating systems alone may challenge the total bandwidth available to the vessel, resulting in patches not being installed when required.

2.2 OPERATIONAL CONSIDERATIONS

The role of seafarers has changed significantly in recent decades, with automation playing an ever-increasing role in day-to-day operations. While this has increased efficiency, offering a steady reduction in the number of people required to operate a vessel, there are other trade-offs that need to be considered, particularly in the context of a geographically distributed virtual team. The current generation of ships are inadvertently provided with remote team members performing tasks such as system monitoring, performance optimisation, software patch installs and in some cases system control [16].

Physical maintenance of technology systems is often performed in port by technicians that may not understand the history of the vessels systems and who will be operating under the vessels tight schedule, leaving limited time for testing and fine tuning of systems.

Controlling and authorising changes made to technology either as configuration changes or software updates are critical to the reliability of the systems. Supporting this and equally important is the capability for centralised management of problems and incidents [5]. While developing these processes and artefacts may be seen as a significant outlay in time and resources, there are opportunities for return on investment where system downtime can be reduced or removed as a result.

2.3 TEAMWORK

Every team requires a number of elements to make it effective - a clearly understood shared purpose, leadership and defined roles for team members. [15] has identified that challenges currently exist within the virtual team due to the lack of a central point of oversight. Off-ship support staff may also be impacted by time zone differences. The effectiveness of a fragmented workforce must also be considered in maintenance / breakdown repair activities and most critically in response to an emergency situation. Where somewhat tenuous communications links are relied upon for team coordination, consideration must be given to process that defines actions that are taken in the event of loss of team communications.

2.4 MANNING

The manning levels needed to operate a vessel safely are defined through a process known as MSM (Minimum Safe Manning). This process requires consideration of risks associated with the sudden failure of critical equipment [15]. When consideration is given to the reduction of crew, the use of shore based resources as part of a virtual team and the increasing role of automation, it is necessary to ask whether it is realistic to revert to manual methods in the event of a broad failure of on-board technology. A scenario such as the Slammer Worm, that was released in 2003 has the potential to simultaneously impact multiple critical systems, as happened in the case of the Davis-Besse power plant in Oak Harbour, Ohio [4]. In the case of a virtual team, relying upon digital communications to collaborate, a broad technology failure may disrupt communications, resulting in additional challenges in resolving such an incident.

2.5 SAFETY, SKILLS & TRAINING

The use of remote support by vendors delivers significant benefits, as they are able to provide highly specialised skills and utilise them across multiple vessels. The skills and training that exists within a team clearly impacts the ability of that team to work effectively. In the case of maritime technology systems, the strategies of relocation of skills to shore and the outsourcing of capabilities to external companies that provide the service remotely need to be monitored to ensure that the necessary skills are maintained within the broader team in a sustainable manner that transcends staff turnovers within service provider companies [16]. It is also necessary to determine how accountabilities that would have traditionally resided with licensed officers can be allocated to shore based organisations [15]. The resultant de-skilling of team members also needs to be considered. Automation has the potential to impact the ability for team members to perform effective problem solving when the automation fails. This is particularly problematic during emergency situations where manual tasks that would normally be performed by automation must be re-configured and shared across team members [15].

3 DISCUSSION

The use of computing technology on ships has changed dramatically in recent years. There has been a transition from proprietary electronics to the use of general purpose computing infrastructure [10]. These systems have significantly different requirements for maintenance and governance. The following sections will discuss the desired changes to increase the maturity of technology governance.

3.1 CHANGE OF PHILOSOPHY FOR TECHNOLOGY ON SHIPS

Keeping technology systems operational includes ensuring that problems and incidents receive appropriate responses; user accounts and access permissions are managed; any changes to configuration or software upgrades are tested and authorised before being applied to the production systems; systems have sufficient capacity and are secured against cyber attack. Planning must also be performed to allow for continuity of operations in the event of a system failure that cannot immediately be resolved. This level of robust management of systems should be initiated at the start of a vessels life and be a constant until it is decommissioned. It may be necessary for the systems and processes to transition through a number of owners and management arrangements during that time so that the necessary technology governance can be maintained.

To reach the target state of maturity and robustness, it is necessary to change the current philosophy of design and operation. Equivalent rigour should be applied to the technology infrastructure, as is found in other aspects of vessel operations. Examples include: the requirement for a 'technology log book' that records significant events that are applicable to the technology systems, as has been required for deck and engine room operations throughout the history of shipping; applying a permit to work scheme for technology changes, requiring vessels to undergo periodic technology surveys and inclusion of technology systems in the planned maintenance schedule.

3.2 FUTURE DESIGN CONSIDERATIONS

System design plays a key part of any robust technological system. When considering that the overall ship IT system must be operated and maintained by a virtual team, this requirement must be considered up front to ensure that appropriate collaboration and governance can occur throughout the full life-cycle of the systems from vessel concept to decommissioning [14]. The design of vessel systems must satisfy the needs for dependability and reliability and enable both onboard and remote operators to work safely, securely and collaborate effectively [15]. Also of critical importance is the requirement for fast recovery of a failed system to restore safety critical operations [11, p.9]. Any design (or modification) of a vessel's technology during its service life needs to consider the full set of operational constraints of the technology services that are provided [13]. This may be documented as a set of 'use cases' that detail the functional tasks that the system provides and shows which 'actors' within the virtual team perform which tasks and any collaboration that is necessary between the actors should be specified. This philosophy may require modification to current practices of developing design specifications and commercial contracts [16].

3.3 THE FUTURE OF TECHNOLOGY GOVERNANCE AND REGULATION

As the policy maker for the maritime industry, the IMO has identified, through the release of their cyber risk management guidelines [9], that risk management for technology systems is fundamental to safe and secure shipping. By 2021 all vessels must include cyber risk management within their safety certificates. There is also an opportunity within the ISPS Code for future inclusion of digital systems and infrastructure within its safety management practices [1]. Lloyds Register has published a ShipRight procedure [14], endorsing a total system approach including consideration of the ship, communications links and off-ship facilities and services that could potentially compromise the safety or the capability of the vessel. This shows an increasing awareness and commitment from key industry organisation that needs to be broadened and translated into operational maturity and governance.

The remediation plans outlined within the ISM and ISPS Codes are required to be reviewed annually, following an incident or after any changes to infrastructure. This ensures that the plans take into consideration the rapidly changing technological solutions offered to the industry. It also enforces the adaptability of these assessments, as a company must prove post-incident that risks are being mitigated.

The approach to providing a more robust technology environment which has been adopted in shore based industries is to implement enterprise capabilities that support the primary mission oriented systems. This approach works well within large enterprises, as it provides consistency and economies of scale such that specialist staff can be employed to provide these services. Using the example of software asset management [15], a large enterprise may deploy a dedicated system that monitors all other infrastructure and periodically gathers data on installed software packages and versions. This is a powerful tool, as it allows for centralised analytics, reporting and identification of non-compliant, vulnerable software. While this approach works well in this type of organisation, enterprise solutions typically require their own specialist resources, infrastructure and connectivity to the supported systems. Enterprise solutions also typically provide services such as identity and access management, configuration management, asset management and centralised management of a Standard Operating Environment (SOE) for desktops and Servers [4]. Maritime based technology operates on a mobile platform with limited connectivity and these arrangements may be challenging to implement in this context.

IT Service Management has been identified by Cater-Steel, Toleman and Tan [5] as a method for providing additional rigour around IT services and reducing the level of ad-hoc operations. An IT Service Management framework such as the IT Infrastructure Library (ITIL) could be applied to the technology services that support operations on sea going vessels. At the tactical level, clear definitions of the various services being provided by technology and their criticality would provide a valuable starting point to stakeholders to understand the technology landscape and the roles that various team members perform in providing those services. An understanding of the current and projected capacities of the infrastructure and realistic plans for maintaining availability / service continuity would complement the service definitions, providing a governance view of the technology as a whole.

3.4 THE VIRTUAL TEAM

The challenge for the virtual team in maintaining sea going technology has many aspects. It includes collaborating with other team members across organisations and geographies, traversing diverse written and spoken languages to support technology infrastructure that is mobile, adding also the difficulties of uncertain connectivity, bandwidth and skills on-board and ashore. It is important to define how much technology work can be managed shore side and what will be done by seafarers, defining boundaries of responsibilities/ dynamics / communications between ship and shore. This is critical to ensure that trust is established between on ship and off ship team members to ensure that the needs of all the team members are being considered [16]. Moving to more robustly governed technology requires the establishment of system boundaries and clear roles / accountabilities (including for any systems hosted or managed ashore). These established boundaries will clearly define any remote connections from ship to shore. There is a need to evolve current management practices to include testing and authorisation of any changes to software or configuration, with roll-back capability in the event of failed changes [14].

4 CONCLUSIONS

Technology has inevitably found its way onto sea going vessels in increasing ways. However, the importance of technology to the operation of the vessel is not represented in the culture of shipping at present. Accountability and ownership for technology is not clearly defined on board. Other critical operations on-board such as navigation and propulsion have formal processes supporting them such as defined roles, accountabilities and the use of logbooks for recording events, the onboard technology is not operated with the same rigour. Similarly, the vessels maintenance program does not typically include details of technology hardware or software patches or software version updates that are required.

Current practices employed for management of technology on sea going vessels has been described as ad-hoc, largely due to the inherent constraints of the operating environment. In addition to this, there is often a lack of technology governance or formally defined IT Service management applied to shipboard technology. Observations of the current industry technology strategy suggest that the levels of automation and the use of 'virtual teams' to support ship based technology will continue to grow. Ship based technology needs to be managed throughout the life-cycle of the vessel and therefore must be considered from the development of the concept, through signing of contracts, build and operation. As the industry moves toward remote operated systems and further reduction of crews, the governance of technology systems will become more critical, as there will be less opportunity for human intervention in the event of a failure of technology at sea.

While there are significant upsides in providing shore based skills to support seafarers (such as sharing extremely specialised skills across a broad range of vessels) there are risks that need to be managed, including de-skilling of seafarers, and lack of ship to shore connectivity during an emergency making shore based support unavailable. The use of virtual teams also presents an organisational and management challenge due to the fact that the members may operate in different geographies, time-zones and organisations. To be successful any well organised team must have a structure, reporting lines and the necessary collaboration tools to ensure that they can achieve their common purpose of supporting the on-board technology systems. The principles of good governance require that any changes made to technology should be authorised by an allocated system owner, who is accountable for the system and its reliable operation.

The lifecycle of technology is much shorter than that of mechanically engineered systems such as a ship. While ships' main and auxiliary systems may be expected to last (with necessary maintenance) for the life of the vessel, technology system lifecycles are much shorter that other equipment on board. Technology requires not only regular maintenance but will also become redundant within a shorter timeframe, requiring technology refit to maintain reliable service. System design is critical when delivering technology systems that must be relied upon. Design does not simply refer to the technology. When considering the entirety of a ship based technology system and the virtual team supporting it the system boundary includes the team members that operate the system and all the enabling processes that occur to support the technology.

Taking the opportunity to adopt a formal ITSM framework such as ITIL to provide governance over the critical technology service on vessels would lift the maturity of the technology services, providing accountability and a reduction in unplanned technology failures.

5 ACKNOWLEDGEMENTS

Rory Hopcraft receives funding from the EPSRC.

6 REFERENCES

[1] A G Bermejo. Maritime Cybersecurity Using ISPS and ISM Codes, 2015.

[2] AMMITEC. Maritime Software Quality Guidelines. 2017.

[3] BIMCO. The guidelines on cybersecurity onboard ships. 2016.

[4] A. Cardenas, S. Amin, B. Sinopoli, A. Giani, A. Perrig, and S. Sastry. Challenges for securing cyber physical systems. In *Workshop on future directions in cyber-physical systems security*, volume 5, 2009.

[5] A. Cater-Steel, M. Toleman, and W.-G. Tan. Transforming it service management-the ITIL impact. *ACIS 2006 Proceedings*, page 81, 2006.

[6] DNV-GL. Recommended practice dnvgl-rp-0496: Cyber security resilience management for ships and mobile offshore units in operation, 2016.

[7] H.B. Esmailia, H. Gardesh, S.S. Sikari. Validating ITIL Maturity to Strategic Business-IT Alignment. *In Proceedings of the2nd International Conference on Computer Technology and Development,* pages 556-561. IEEE, 2010.

[8] IMO. Report of the Maritime Safety Committee on its Eightieth Session MSC 80/24, 2005.

[9] IMO. Guidelines on Maritime Cyber Risk Management MSC-FAL.1/Circ.3, 2017.

[10] C. W. Johnson. Why we cannot (yet) ensure the cybersecurity of safety-critical systems. 2016.

[11] K. D. Jones, K. Tam, and M. Papadaki. Threats and impacts in maritime cyber security. 2016.

[12] S. K. Katsikas. Cyber security of the autonomous ship. In *Proceedings of the 3rd ACM Workshop on Cyber-Physical System Security*, pages 55–56. ACM, 2017.

[13] A. Kott and I. Linkov. Cyber resilience of systems and networks, 2018.

[14] Lloyds Register. Cyber-enabled ships: Deploying information and communications technology in shipping, Lloyds Register's approach to assurance. 2016.

[15] H. A. Oltedal and M. Lützhöft. *Managing Maritime Safety*. Routledge, 2018.

[16] B. Twomey. The cyber-enabled ship what are the risks and what are the mitigations? *Encyclopedia of Maritime and Offshore Engineering*, pages 1–17, 2017.

7 AUTHORS BIOGRAPHY

Gerd Berner is a PhD Researcher from University of Tasmania (AMC). His research is focused on cyber security resilience of maritime technology.

Rory Hopcraft is a PhD Researcher from Royal Holloway University of London. His research is focused on regulatory aspects of maritime cyber security.

Joel Scanlan is a lecturer in ICT at the University of Tasmania. His research background and primary teaching area are in Cyber Security.

Margareta Lutzhoft is a Professor in the MarSafe group at the Western Norway University of Applied Sciences. Her research interests include human-centered design and the effects of new technology.

Jonathan Earthy is a Principal Human Factors Specialist for Lloyd's Register Marine and Offshore. He coordinates standards, research and assessment activities and products. His specialist area is assurance of the quality of Human Factors work, in particular for complex IT systems.

PRESENTATION

APPLICATION OF FUZZY COGNITIVE MAPS TO INVESTIGATE THE CONTRIBUTORS OF MARITIME GROUNDING ACCIDENTS

B Navas de Maya* and **R E Kurt** University of Strathclyde, UK

SUMMARY

The maritime sector is often criticised for being reactive in safety where safety measures are considered after major accidents. However, a new safety approach is becoming popular which focuses not only what goes wrong but also what goes right. This approach allows us to identify required abilities from crew for resilient shipping operations. In order to better identify these necessary resilience abilities, this paper aims to assess the factors affecting grounding accidents by using Fuzzy Cognitive Maps (FCMs), which calculates and evaluates the individual weights of the underlying factors involved in grounding accidents. FCMs appear to be a suitable approach since it can model complex relationships between these contributing factors. Therefore, this study is focused on past accidents on general cargo vessels. General cargo vessels were selected for this study as information was provided regarding 122 accidents in this vessel category. Then, the Marine Accident Investigation Branch (MAIB) database was analysed to obtain the weightings for the contributors in grounding accidents, which help us to understand in which areas resilience abilities are required the most. The results indicate that Fatigue is the highest contributor into grounding accidents in general cargo vessels.

NOMENCLATURE

EMCIP	European Marine Casualty Information Platform
FCMs	Fuzzy Cognitive Maps
MAIB	Marine Accident Investigation Branch

1. INTRODUCTION

Historic accident analysis show that shipping accidents have always been a concern for the maritime sector [1]. In order to overcome these accidents, the shipping industry has implemented various measures in the past decade, which overall aim to improve maritime safety. However, despite all the efforts, accidents are still happening and they remain a major concern when around 90% of world trading is still carried out by shipping companies [2]. Hence, efforts must be focused on a reduction in the accident rate through the implementation of new safety measures.

When analysing the literature available, human factors are identified as the major cause of accidents not only in maritime but also in various strategic sectors such as nuclear, aviation or aerospace [3]. For instance, human errors are responsible for about 70-80% of the accidents in the aviation sector [4], while they represent more than 80% of accidents for the shipping companies [5].

With the purpose of identifying and understanding human factors implication into past accidents, this paper performs a detailed analysis of an historical accident database. In order to fulfil this objective, an historical accident database is provided by the Marine Accident Investigation Branch (MAIB). MAIB database includes marine accidents involving UK vessels worldwide and all vessels operating in UK territorial water for the period 2000-2011. Figure 1 shows the accidents classification regarding the accident category.

The main problem related with MAIB database is the availability of duplicate causalities in addition to partial missing information. Hence, for this study only 2690 accidents will be further considered, as a complete description is provided for these accidents, including human and technical factors.

It can be seen from Figure 1 that 'hazardous incident', 'accident to person' and 'machinery failure' are the most populated accident categories. However, in this study 'grounding' is selected to be investigated as the consequences of grounding have potential to impact on the entire vessel and surrounding environment. A grounding accident may be defined as an accident in which the vessel makes involuntary contact with the ground, excluding briefly contact where no damage to the ship is caused.

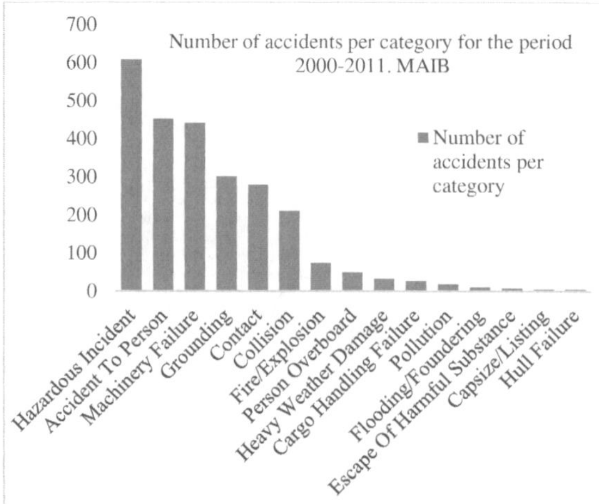

Figure 1: Number of accidents per category provided by MAIB for the period 2000-2011. 2690 accidents are considered fort his study.

In the next sections of this paper a FCM will be developed with the aim to show how human and technical factors, contribute to maritime grounding accidents in general cargo vessels.

2. METHODOLOGY

The aim of this paper is to model a FCM that will represent the interrelation between human and technical factors involved within grounding accidents in general cargo vessels. In order to achieve this objective, the following steps are followed: Firstly, the FCM method is introduced and explained briefly. Secondly, the vessel categories and the factors that are responsible for grounding accidents are identified and named according with MAIB guidelines. Also the reasons for selecting general cargo vessels are highlighted. It is important to mention that these factors have been established and predefined by the MAIB, hence, the nomenclature which refers to human factors may vary from other sources in the literature. Moreover, the process to obtain each component of a FCM, i.e. interaction matrix and state vector, is explained. Furthermore, the threshold function is applied, and a complete FCM for general cargo vessels is represented, in order to show the overall process. Additionally, the result section shows the final weight distribution of factors in grounding accidents obtained for general cargo vessels.

2.1 FUZZY COGNITIVE MAPS (FCMS)

Maritime accidents are often defined as complex processes caused by an interaction of various factors. Hence, normally there is no a simple solution to prevent them [6]. Due to the vagueness and data unavailability when an accident occurs, and that this study analyses data collected from past experiences, a method that can deal with both as the FCM method [7] is the most suitable option for this study.

FCMs were developed in 1986 as cognitive map in which the concepts are weighted with fuzzy numbers [8]. According to the available literature there are many definition regarding FCM. For instance, [9] defines a FCM as a description of the behavior of a system through concepts, in which each concept represents an entity, a state, or a characteristic of the system. Moreover, FCMs may be defined as extensions of cognitive maps which model complex chains of casual relationships amongst factors [10].

2.1 (a) Mathematical Representation of a FCM

FCMs can be obtained through three main components. Firstly, an interaction matrix with dimension n x n where n indicates the number of concepts represented in the FCM. Moreover, a zero value in the matrix indicate that a relation does not exist between two particular elements, while non-zero elements show not only that there is a relation between two elements but also the strength of that relation. Figure 2 shows an example of a simple FCM and its equivalent transition matrix. Secondly, an initial state vector, which shows the initial value of the concepts in the scenario being modelled at any point in time (t) before applying the threshold function. Finally, a threshold function. Although there are plenty threshold functions available [11], the Sigmoid function gives any possible value within the interval [0,1], [12] and it has been proved that using this function provides greater benefits [13]. Hence, the Sigmoid function is selected for this study as shown in Equation 1.

$$A_i^{(t+1)} = \frac{1}{1+e^{-\lambda x}} \quad (1)$$

Within a FCM, each of the concepts is represented by a number, A_i, that represents its value within the interval [0,1] [14], which assures that it is possible to identify three types of connections between the concepts described in the FCM that represents the nature of their respective influence:

- A positive value between the weights of factors C_i and C_j ($W_{ij}>0$). Hence, an increase in the first factor will lead to an increase in the second factor. Moreover, if the first factor is decreased the second factor will be also decreased.
- A negative value between the weights of factors C_i and C_j ($W_{ij}<0$). Then, an increase in the first factor will lead to a decrease in the second factor, while if the first factor is decreased the second factor will be increased. However, since we are investigating the factors that are contributing to the development of accidents, we do not have an example of a negative weighting in this study.
- No causality ($W_{ij}=0$) which indicates that there is no relation between two factors.

According with [8], a traditional formula to calculate the values of the concepts in a FCM is as shown in Equation 2:

$$A_i^{(t+1)} = f\left(A_i^{(t)} + \sum_{j=1, j\neq 1}^{n} W_{ji} A_j^{(t)}\right) \quad (2)$$

In which $A_i^{(t+1)}$ represents the value of the factor C_i at the step t+1, f is the threshold function which will make sure the concept value is limited within the interval [0,1], W_{ji} represents the weight between both concepts C_i and C_j, and $A_j^{(t)}$ is the value of the concept C_j at step t.

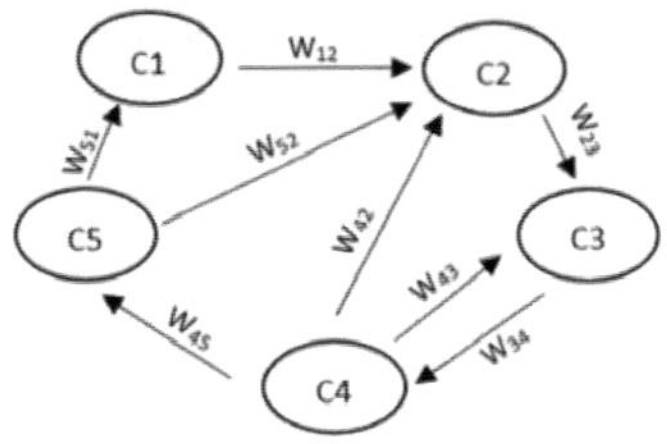

$$\begin{bmatrix} 0 & w12 & 0 & 0 & 0 \\ 0 & 0 & w23 & 0 & 0 \\ 0 & 0 & 0 & w34 & 0 \\ 0 & w42 & w43 & 0 & w45 \\ w51 & w52 & 0 & 0 & 0 \end{bmatrix}$$

Figure 2: A simple representation of a FCM and its equivalent transition matrix [15].

In order to create a FCM, Equation 2 is applied as an iterative process for each time step (step 1, step 2 etc.) until the process ends, which could happen in three different scenarios [12]:

- The FCM reaches equilibrium. After two consecutive iterations the results are identical. In this case the simulation stops and the FCM is considered steady.
- The FCM does not produce a stable result. The results keep cycling between a set of values without stabilizing. This situation is known as the "limit cycle", and it results from a certain combination of weight values when applying a FCM, which drive the map away from reaching equilibrium [16].
- A last possibility occurs when the FCM does not reach identical values, producing different state vectors for each step. This possibility is defined as 'chaos' and it can appear in complex scenarios.

This FCM method has been applied on collision accidents in the past [15], showing promising results for collision and that this method is working effectively. Hence, it is applied in this paper in order to obtain the weights of human and technical factors for grounding accidents in general cargo vessels.

2.2 HUMAN AND TECHNICAL FACTORS AFFECTING GROUNDING ACCIDENTS

As it was aforementioned, the factors that are considered within this study are obtained from analyzing a database provided by the Marine Accident Investigation Branch (MAIB) for the period 2000-2011. This period was selected for this study due to the fact that MAIB changed its accident taxonomy in 2011, in order to comply with the European Marine Casualty Information Platform (EMCIP) taxonomy. According with the database, there are eight vessel categories which have suffer from at least one grounding accidents within the period 2000-2011. Figure 3 shows the distribution of grounding accidents amongst the vessel categories. The first column for each vessel category shows the total number of accidents. However, not all of these accidents provide information regarding the human and technical factors that leaded to them. Then, the second columns for each vessel category shows the number of accidents which contains data related to these human/technical factors. As it can be observed from the figure, 'General cargo' category shows the highest number of accident in both columns. Thus, general cargo is selected for this study.

Furthermore, the database for grounding accidents in general cargo is composed of 65 factors, both human and technical as shown in Table 1.

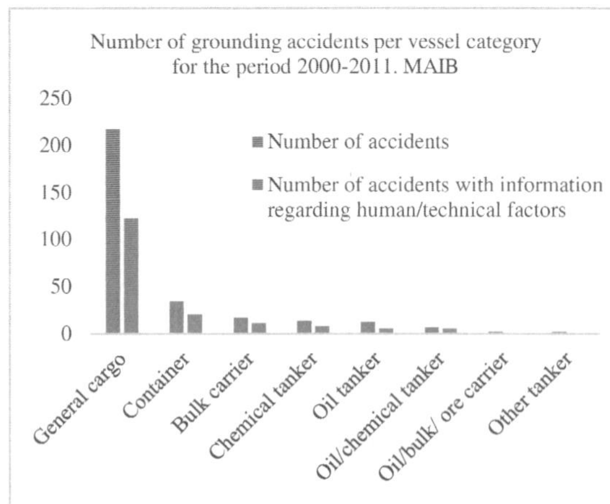

Figure 3: Grounding accidents amongst various vessel categories provided by MAIB for the period 2000-2011. 122 accidents in general cargo are considered for this study.

HUMAN AND TECHNICAL FACTORS IN GROUNDING ACCIDENTS		
Factor 3 – Alcohol use	Factor 38 – Heavy weather	Factor 75 - Poor regulations, policies or practices
Factor 4 – Allocation of responsibility inappropriate	Factor 41 - Inadequate management of physical resources	Factor 76 - Pressures - organisational
Factor 5 – Assembly defect	Factor 42 - Inadequate manning	Factor 77 - Procedures inadequate
Factor 6 – Automation means crew not trained in use of manual alternatives	Factor 44 - Inadequate resources	Factor 80 – Safety culture
Factor 9 – Characteristic defect	Factor 45 - Inattention	Factor 84 - Situational awareness or communication inadequate
Factor 10 - Communication	Factor 48 - Knowledge of regulations/standards	Factor 87 – System defect

HUMAN AND TECHNICAL FACTORS IN GROUNDING ACCIDENTS		
	inadequate	
Factor 11 - Communication or language difficulties	Factor 49 - Knowledge of ship operations inadequate	Factor 89 - Task difficulty
Factor 12 - Company standing orders inadequate, insufficient, conflicting	Factor 50 - Lack of communication or co-ordination	Factor 90 - Technical knowledge inadequate
Factor 13 - Competence	Factor 51 – Lack of self-discipline	Factor 91 - Temperature
Factor 14 - Complacency	Factor 52 – Language problem	Factor 92 - Time pressures
Factor 16 - Corrosion	Factor 55 - Management and supervision inadequate	Factor 94 - Training which itself is inadequate
Factor 17 - Culture	Factor 56 - Manning (rotation /watches)	Factor 95 - Training, inexperience, knowledge
Factor 18 - Current	Factor 59 - Misapplication of regulations, policies, procedures or practices	Factor 96 - Training, skills, knowledge
Factor 19 – Design inadequate	Factor 62 – Operation Instructions inadequate	Factor 99 - Unaware of role/task/responsibility
Factor 24 – Equipment misuse	Factor 63 – Other vessel	Factor 100 - Uncharted underwater Obstruction
Factor 25 - Equipment not available	Factor 65 - Perception abilities	Factor 101 - Under stimulation
Factor 26 - Equipment poorly designed for operational use	Factor 66 – Perception of Risk	Factor 104 - Vigilance
Factor 28 - Excessive work load	Factor 67 – Personal problem	Factor 105 - Violation of procedures
Factor 30 - Fatigue	Factor 68 – Personality	Factor 106 – Visibility
Factor 31 - Fatigue and vigilance	Factor 69 - Personnel unfamiliar with equipment/not trained in use	Factor 107 – Visual environment
Factor 35 - Hazardous natural environment	Factor 71 - Poor decision making/information use	Factor 110 – Worn out
Factor 36 - Health: drugs/alcohol	Factor 72 - Poor housekeeping	

Table 1: Human and technical factors involved in grounding accidents in general cargo vessels.

2.3 COMPONENTS IN A FCM

Once the factors involved in grounding accidents in general cargo vessels were defined, a FCM was created. As it was aforementioned, a FCM requires three components to be created:

- First, an interaction matrix with dimension n x n where n indicates the number of concepts analyzed in the FCM,
- second, an initial state vector, which displays the initial value of the concepts in the scenario being modelled at any point in time (t)
- and at last, a threshold function. As it was explained above, for this study the Sigmoid function was selected.

2.3 (a) Interaction Matrix

When in real life experts are asked to provide their opinion, it is due to the fact that they have accumulated knowledge about a specific problem based on their past experiences [17]. However, this action also incorporates a subjective element due to a lack of knowledge from some experts. On the other hand, by considering and analyzing a database as if it was an expert, it is possible to obtain fully objective results, as the accidents analyzed have already happened and it is possible to track back the factors that leaded to them.

Then, in order to fulfill the aim of this paper, MAIB historical accidents database was analyzed, comparing each pair of accident factors. For example, in order to determinate the relation amongst *Factor 13 – Competence* and *Factor 14 – Complacency* mentioned in Table 1, the accident database was filtered by the accidents that any of these two factors caused. Then, the number of accidents was filtered again by the accidents that shared both factors as a common accident cause.

Following this process, the influence of *Factor 13 – Competence* over *Factor 14 – Complacency* was obtained by looking into the relation between the accidents with both factors in common and the accidents with *Factor 13 – Competence* but not *Factor 14 – Complacency*. In the same way, the weight of *Factor 14 – Complacency* over *Factor 13 – Competence* is the relation between the accidents with both factors presented and the accidents with *Factor 14 – Complacency* but not *Factor 13 – Competence*. This process is represented in Equation 3 as follow:

$$W_{F13-F14} = \frac{w_{F13\cap F14}}{w_{F13\setminus F14\,(set\ subtraction)}} \tag{3}$$

The same process is repeated in order to obtain the relations and weights of each pair of factors. Table 2 shows a partial representation of the interaction matrix for grounding accidents in general cargo vessels for the period 2000-2011. Moreover, due to size limitations the interaction matrix is just partially represented in Table 2.

	F3	...	F12	F13	F14	F15	F16	...
F3	0.00	...	0.00	0.25	0.00	0.00	0.00	...
...	...	...	...	...	...	...	...	...
F12	0.00	...	0.00	0.27	0.36	0.00	0.00	...
F13	0.07	...	0.20	0.00	0.07	0.00	0.00	...
F14	0.00	...	0.40	0.10	0.00	0.00	0.00	...
F15	0.00	...	0.00	0.00	0.00	0.00	0.00	...
F16	0.00	...	0.00	0.00	0.00	0.00	0.00	...
F17	0.00	...	0.00	0.00	1.00	0.00	0.00	...
...	...	...	...	...	...	...	...	...

Table 2: Partial representation of the interaction matrix for grounding accidents in general cargo vessels. Period 2000-2011.

2.3 (b) State Vector

Regarding the state vector, it was defined as the percentage of each factor within the total number of accidents in general cargo vessels. Table 3 shows as an example the state vector for grounding accidents in general cargo vessels for the period 2000-2011.

F3	**F4**	**F5**	**F6**	**F9**	**F10**
0.03	0.01	0.01	0.01	0.02	0.01
F11	**F12**	**F13**	**F14**	**F16**	**F17**
0.01	0.09	0.12	0.08	0.01	0.01
F18	**F19**	**F24**	**F25**	**F26**	**F28**
0.12	0.02	0.02	0.01	0.01	0.02
F30	**F31**	**F35**	**F36**	**F38**	**F41**
0.09	0.02	0.02	0.01	0.02	0.04
F42	**F44**	**F45**	**F48**	**F49**	**F50**
0.01	0.03	0.09	0.01	0.05	0.07
F51	**F52**	**F55**	**F56**	**F59**	**F62**
0.02	0.01	0.01	0.02	0.07	0.02
F63	**F65**	**F66**	**F67**	**F68**	**F69**
0.02	0.03	0.02	0.01	0.02	0.04
F71	**F72**	**F75**	**F76**	**F77**	**F80**
0.05	0.01	0.02	0.03	0.16	0.03
F84	**F87**	**F89**	**F90**	**F91**	**F92**
0.19	0.02	0.02	0.05	0.01	0.01
F94	**F95**	**F96**	**F99**	**F100**	**F101**
0.04	0.02	0.01	0.01	0.12	0.02
F104	**F105**	**F106**	**F107**	**F110**	
0.02	0.02	0.01	0.01	0.01	

Table 3: State vector for grounding accidents in general cargo vessels. Period 2000-2011.

2.3 (c) FCM Representation

As it was explained previously, this paper analyses historical accident data obtained from MAIB for the period 2000-2011. After analyzing the database, 65 factors, mentioned in Table 1, were identified as a major cause in at least one accident in general cargo vessels. The FCM was created following Equation 2 until it reached equilibrium. As an example to illustrate this process, Figures 4 to 7 shows the variation in the weights without normalize for both human and technical elements involved in grounding accidents in general cargo vessels until equilibrium is reached, which occurs before step 6 for this example.

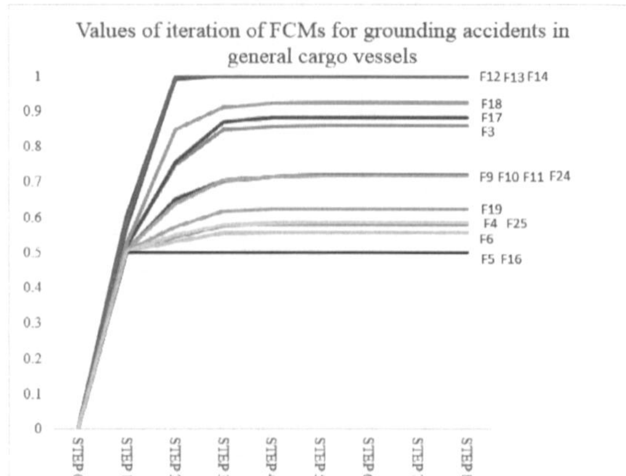

Figure 4: Values of FCM for grounding accidents in general cargo vessels until equilibrium is reach for factors 3-25. Period 2000-2011.

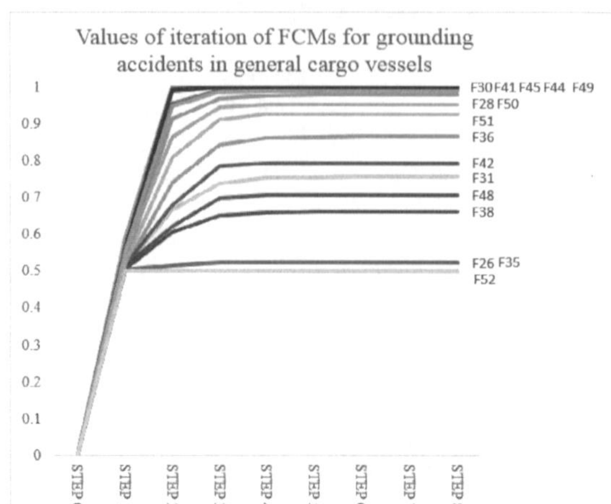

Figure 5: Values of FCM for grounding accidents in general cargo vessels until equilibrium is reach for factors 26-52. Period 2000-2011.

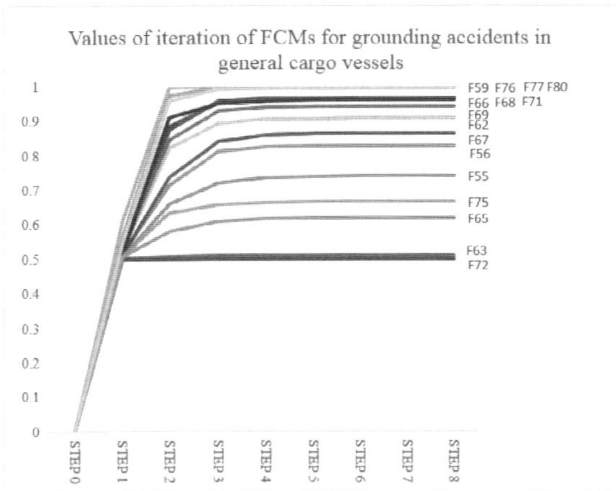

Figure 6: Values of FCM for grounding accidents in general cargo vessels until equilibrium is reach for factors 55-80. Period 2000-2011.

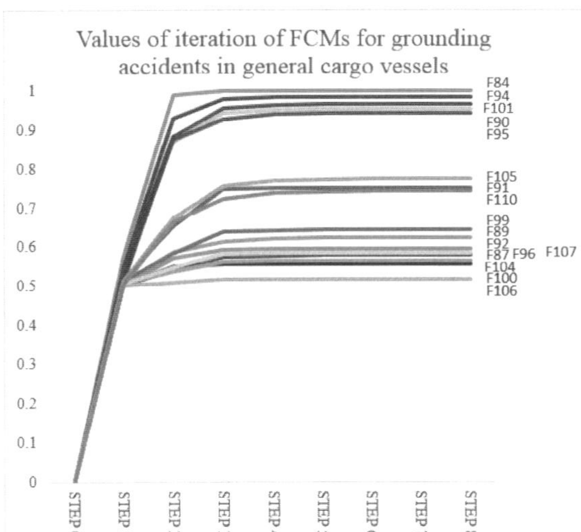

Figure 7: Values of FCM for grounding accidents in general cargo vessels until equilibrium is reach for factors 84-110. Period 2000-2011.

3. RESULTS AND DISCUSSION

Finally, after creating a FCM, the results were normalized and the final weights of grounding accidents contributors for general cargo vessels were obtained and ranked. Table 4 shows the weights of contributors for grounding accidents in general cargo vessels.

Analyzing the results obtained from Table 4, it is clearly shown that there are 65 factors that lead to grounding accidents. Furthermore, from all the previous factors, twelve are not related with human factors, leading to 15.67% of accidents, which reinforces the theory regarding human factors as the main cause in maritime accidents [5].

Moreover, according to Table 4, *Factor 30 – Fatigue* has the highest influence while *Factor 72 – Poor housekeeping* has the minimum impact for grounding. These factors have also been normalized in Table 4, in order to show how each human/technical factor contribute into accidents in terms of %.

Fatigue is widely recognized as a hazard for safety not only in the maritime sector [18] but also in other sectors e.g. aviation or aerospace [19]. Hence, it is not surprising the fact that this factor has the highest contribution into grounding accidents in general cargo vessels.

General cargo			
Factor ranking	**Factor description**	**Weight from FCM**	**Weight normalized (%)**
30	Fatigue	1.000	2.104
77	Procedures inadequate	0.999	2.102
12	Company standing orders inadequate, insufficient, conflicting	0.998	2.100
14	Complacency	0.997	2.098
45	Inattention	0.993	2.089
13	Competence	0.991	2.087
41	Inadequate management of physical resources	0.988	2.079
84	Situational awareness or communication inadequate	0.986	2.075
76	Pressures - organisational	0.974	2.051
59	Misapplication of regulations, policies, procedures or practices	0.973	2.048
80	Safety culture	0.959	2.019
44	Inadequate resources	0.955	2.010
49	Knowledge of ship operations inadequate	0.945	1.989
94	Training which itself is inadequate	0.927	1.950
28	Excessive work load	0.913	1.921
71	Poor decision making/information use	0.911	1.918
66	Perception of risk	0.887	1.866
101	Understimulation	0.882	1.856
68	Personality	0.875	1.842
95	Training, inexperience, knowledge	0.874	1.840
90	Technical knowledge inadequate	0.869	1.830
50	Lack of communication or co-ordination	0.864	1.818
18	Current	0.847	1.783
69	Personnel unfamiliar with equipment/not trained in use	0.846	1.781
62	Operating instructions inadequate	0.822	1.731
51	Lack of self-discipline	0.808	1.701
17	Culture	0.755	1.590
3	Alcohol use	0.747	1.573

36	Health: drugs/alcohol	0.740	1.556
67	Personal problem	0.740	1.556
56	Manning (rotation /watches)	0.715	1.505
42	Inadequate manning	0.680	1.430
105	Violation of procedures	0.673	1.417
31	Fatigue and vigilance	0.667	1.403
55	Management and supervision inadequate	0.662	1.394
110	Worn out	0.662	1.394
10	Communication	0.652	1.372
11	Communication or language difficulties	0.652	1.372
91	Temperature	0.651	1.371
9	Characteristic defect	0.648	1.363
24	Equipment misuse	0.639	1.344
75	Poor regulations, policies or practices	0.636	1.339
48	Knowledge of regulations/standards inadequate	0.622	1.309
38	Heavy Weather	0.607	1.276
89	Task difficulty	0.586	1.233
99	Unaware of role/task/responsibility	0.582	1.225
65	Perception abilities	0.581	1.223
19	Design Inadequate	0.573	1.206
92	Time pressures	0.570	1.201
25	Equipment not available	0.550	1.157
100	Uncharted underwater Obstruction	0.550	1.157
107	Visual environment	0.548	1.153
4	Allocation of responsibility inappropriate	0.543	1.143
87	System defect	0.543	1.143
96	Training, skills, knowledge	0.542	1.142
104	Vigilance	0.536	1.128
6	Automation means crew not trained in use of manual alternatives	0.531	1.118
35	Hazardous natural environment	0.517	1.088
26	Equipment poorly designed for operational use	0.513	1.081
106	Visibility	0.510	1.072
63	Other Vessel	0.506	1.065
5	Assembly defect	0.500	1.052
16	Corrosion	0.500	1.052
52	Language problem	0.500	1.052
72	Poor housekeeping	0.500	1.052

Table 4: Final weight of contributors for grounding accidents in general cargo vessels. Period 2000-2011.

4. CONCLUSIONS

In this paper, a new modelling and simulation approach based on FCMs was proposed for assessing the factors that lead to grounding accidents in general cargo vessels. The main target that was aimed to achieve was to demonstrate that it is possible to define generic weights for human and technical factors and also that the obtained weights could be applied at any company in order to identify main areas where safety can be increased and therefore accidents could be reduced.

As a result of this work, weights for grounding accidents within general cargo vessels were obtained and displayed in the previous section. As it was shown in the result section, "Fatigue" was obtain as the highest contributor factor into accidents in general cargo vessels.

Furthermore, this weighting table could be used by managers in a company for identifying the main factors that lead to accidents and address them in order to increase safety by decreasing the number of accidents.

In addition, for this study other consideration could have been taken into account, for instance, the data analysed was taken from MAIB database between 2000 and 2011, as MAIB has changed its accidents' nomenclature from 2011. However, a more up-to-date database could be analysed to compare if the factors that caused accidents in the past have been addressed thought safety measures or if they are still leading to grounding accidents nowadays. Moreover, this method could be applied in other sectors, i.e. aviation, in order to compare how human factors causing accidents differ between transport modes.

5. ACKNOWLEDGEMENTS

The authors of this paper would like to acknowledge the assistance of the Marine Accident Investigation Branch (MAIB) for providing the required data for performing this study.

6. REFERENCES

1. ELIOPOULOU, E., PAPANIKOLAU, A. & VOULGARELLIS, M., Statistical analysis of ship accidents and review of safety level', *Safety Science, 85, 282-292,* 2016.

2. CHAUVIN, C., LARDJANE, S., MOREL, G., CLOSTERMANN, J.-P. & LANGARD, B., 'Human and organisational factors in maritime accidents: Analysis of collisions at sea using the HFACS', *Accident Analysis & Prevention, 59, 26-37,* 2013.

3. AZADEH, A. & ZARRIN, M., 'An intelligent framework for productivity assessment and

analysis of human resource from resilience engineering, motivational factors, HSE and ergonomics perspectives', *Safety Science, 89, 55-71,* 2016.

4. O'HARE, D., WIGGINS, M., BATT, R. & MORRISON, D., 'Cognitive failure analysis for aircraft accident investigation', *Ergonomics, 37, 1855-1869,* 1994.

5. TURAN, O., KURT, R. E., ARSLAN, V., SILVAGNI, S., DUCCI, M., LISTON, P., SCHRAAGEN, J. M., FANG, I. & PAPADAKIS, G., 'Can We Learn from Aviation: Safety Enhancements in Transport by Achieving Human Orientated Resilient Shipping Environment', *Transportation Research Procedia, 14, 1669-1678,* 2016.

6. KRISTIANSEN, S, 'Maritime Transportation: Safety Management and Risk Analysis', *Taylor & Francis,* 2013.

7. AZADEH, A., SALEHI, V., ARVAN, M. & DOLATKHAH, M, 'Assessment of resilience engineering factors in high-risk environments by fuzzy cognitive maps: A petrochemical plant', *Safety Science, 68, 99-107,* 2014.

8. KOSKO, B, 'Fuzzy cognitive maps', *International journal of man-machine studies, 24, 65-75,* 1986.

9. XIROGIANNIS, G. & GLYKAS, M, 'Fuzzy cognitive maps in business analysis and performance-driven change', *IEEE Transactions on Engineering Management, 51, 334-351,* 2004.

10. EDEN, C, 'Cognitive mapping', *European Journal of Operational Research, 36, 1-13,* 1988.

11. MOHR, S, 'Software design for a fuzzy cognitive map modeling tool', *Tensselaer Polytechnic Institute,* 1997.

12. XIAO, Z., CHEN, W. & LI, L, 'An integrated FCM and fuzzy soft set for supplier selection problem based on risk evaluation', *Applied Mathematical Modelling, 36, 1444-1454,* 2012.

13. SALMERON, J. L., 'Supporting decision makers with fuzzy cognitive maps', *Research-Technology Management, 52, 53-59,* 2009.

14. LEÓN, M., RODRIGUEZ, C., GARCÍA, M. M., BELLO, R. & VANHOOF, K., 'Fuzzy cognitive maps for modelling complex systems'. *Mexican International Conference on Artificial Intelligence. Springer, 166-174.,* 2010.

15. NAVAS DE MAYA, B., KURT, R. E. & TURAN, O., 'Application of fuzzy cognitive maps to investigate the contributors of maritime collision accidents'. *Proceedings of 7th Transport Research Arena TRA April 16-19, Vienna, Austria,* 2018.

16. WIERZCHON, S. T., 'The fuzzy systems handbook. A practitioner's guide to building, using, and maintaining fuzzy systems'. *Pergamon,* 1995.

17. SHANKAR, A. A., 'Opinion mining for Decision Making in Medical Decision Support system - A Survey-'. *Proc. of the Second International Conference on Computer Applications, ICCA,* 2012.

18. STRAUCH, B., 'Investigating fatigue in marine accident investigations'. *6th International Conference on Applied Human Factors and Ergonomics, AHFE,* 2015.

19. MURRAY, S. L. & THIMGAN, M.S., 'Human fatigue risk management. Improving safety in the chemical processing industry, 2016.

7. AUTHORS BIOGRAPHY

Beatriz Navas de Maya is a PhD researcher in the Naval Architecture, Ocean and Marine Engineering Department at the University of Strathclyde. She started her PhD in October 2016 in the fields of maritime safety, human factors and resilience engineering. Her PhD thesis focuses on how it is possible to enhance safety by considering human factors, resilience engineering concepts and principles as well as the accident investigation data in an integrated manner.

Dr Rafet Emek Kurt is a Lecturer and PGT Recruitment Director in the Department. Dr Kurt's research interests lie in the field of human factors in maritime domain which includes design, optimisation, operation of ships, by considering the human factors, safety and risk at the core of the subject. Dr Kurt also investigated the effects of noise and vibration on human performance and developed human response models for noise and vibration levels on-board ships. Besides his primary research area, Dr Kurt is also actively working in the area of ship recycling. He has contributed to development of tools, models and innovative training for the industry. Dr Kurt was a member of ship recycling working group when Hong Kong Convention was being developed at International Maritime Organisation (IMO). Dr Kurt is actively attending IMO's periodic MSC and MEPC meetings.

PRESENTATION

HUMAN FACTORS INVESTIGATION IN MARITIME INCIDENT ANALYSIS FOR IMPROVED SHIP SAFETY: AN INNOVATIVE APPROACH BASED ON SYSTEM THEORY

C Bongermino and **P Gualeni,** University of Genoa, IT

SUMMARY

The complexity of Human Factors interactions with ship safety will be addressed within the field of maritime accident analysis by means of the Causal Analysis using System Theory (CAST). This is a specific tool within the more comprehensive System-Theoretic Accident Model and Processes (STAMP) approach, already applied in other fields like aviation, defence, healthcare with significant outcomes in terms of safety enhancements.

CAST can provide a framework to understand the entire accident process and identify the most important systemic causal factors, including the systemic socio-technical ones, since it provides the capability to spot weakness in existing safety control structure in order to possibly implement advances.

In this perspective, the application to the Herald of Free Enterprise accident will be discussed, with focus on CAST effectiveness in assessing the complex Human Factors interaction with the ship safety control structure and in providing input to continuously enhance the ship safety and the ship resilience by design.

1. INTRODUCTION

It is well recognized that quite frequently accidents are at the base of significant safety rulemaking activity, also at IMO: SOLAS has been originated as a reaction to Titanic disaster, Torrey Canyon was the decisive event for MARPOL development, Amoco Cadiz was relevant as well for significant MARPOL amendments and for the establishment of STCW Convention, Herald of Free Enterprise is considered the tragedy at the base of ISM Code implementation as mandatory with the introduction of SOLAS chapter IX.

It is worth mentioning that in the latest decades the need to complement this reactive approach with a proactive one has increasingly been tackled as well. Several initiatives are under development at IMO to this aim in order to provide a robust framework to guarantee that the necessary issues are addressed with the adequate priority and stakeholders support, before accidents happen [1, 2, 3].

In any case, a complementary and integrated attitude between reactive and proactive approach is beneficial and definitely accidents might always represent a source of clear indications in setting the priorities for the safety rulemaking agenda.

This paper focuses, in particular, on accident analysis by means of the application of a system-based methodology, innovative within the maritime context.

In the perspective that one of the pillars for sound safety rules development is to learn from accident, it is, in fact, important a continuous evolution of tools that can better support this task. One of the most significant challenges is adequately analyse real accident when the Human Factor is claimed to play a significant role.

In such cases the traditional event chain might be not exhaustive and innovative frameworks are necessary, especially to identify the influence of systemic factors in the whole dynamic of the accident [4]. The analysis process, in fact, should be able to put in evidence the effect of the environmental/operational constraints and the organizational boundary conditions on human behaviour and on the general evolution toward the final loss.

In this paper the Causal Analysis using System Theory (CAST), that is a specific tool within the more comprehensive System-Theoretic Accident Model and Processes (STAMP), will be described and applied to the Herald of Free Enterprise accident. In the investigation activity following this tragedy, in fact, Human Factor, management and organizational issues, together with a number of operational/environmental conditions, have been spotted as significant factors [5, 6].

With the CAST methodology, all these aspects will be logically structured and visualized thanks to a system thinking approach, that will be detailed in the following paragraph, where also the importance of safety constraints and safety control structure will be pointed out.

2. THE CAST METHODOLOGY

Most traditional accidents reports are written from the perspective of an event-based model in order to describe the events and identify one or more of this events as the root cause. However, it is very frequent that the analysis of why those events occurred shouldn't stop after finding someone to blame, usually a human operator, taking therefore the opportunity to learn important lessons on a larger scale.

An accident analysis technique has to provide a framework or process to assist in understanding why the events occurred, i.e. the entire accident process and identifying the most important systemic causal factors involved. In this perspective, an innovative approach to accident analysis is represented by CAST (Causal Analyses using System Theory).

CAST can be used to identify the questions that need to be answered to fully understand why the accident occurred and to maximize learning from the event.

It takes its roots from STAMP (System-Theoretic Accident Model and Processes), developed by Prof. N.

Leveson at the Massachusetts Institute of Technology, based on system thinking applied to safety. In fact, safety is a system property, not a component property and it must be controlled at the system level, not only at the component level [7].

The accident process is documented by showing the socio-technical safety control structure for the system involved and the safety constraints that were violated at each level of this control structure and why. To understand this approach, both concepts of "safety control structure" and "safety constraints" are of outstanding importance.

The framework of CAST is based on a vision of complex systems as a hierarchy of organizational levels. Each level is characterized by emergent properties, resulting from complex interactions among system components, that represent constraints on the degree of freedom of components at the level below. If safety is an emergent property, unsafe system behaviour is defined in terms of "safety constrains" on the behaviour of the system components. Therefore, safety performance is a control problem (enforcing "safety constraints") rather than a failure or reliability problem. Between each level of the hierarchy there is a feed-back control loop acting to ensure that the safety constraints are enforced (Figure 1) [8].

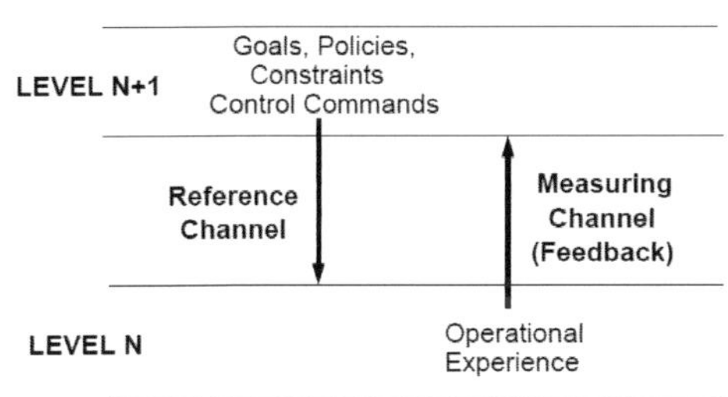

Figure 1: Control channels between levels [7].

A "safety control structure" is the framework where the hierarchy of organizational levels are connected by effective communication channels made of control actions (top-down arrows) and feed-back about control satisfaction (bottom-up arrows).

An effective representation of the relation between accident and inadequate enforcement of safety controls is then given in Figure 2.

The aim of CAST does not consist of identifying single causal factors or chain of events, but it goes deeper providing the capability to examine the entire socio-technical system design to identify the weakness in existing safety control structure and come up with changes that will not simply eliminate symptoms, but potentially all the causal factors, including the systemic ones.

In order to better understand how to face an accident investigation using CAST, the following list of steps has been identified to describe the general process [7]. It is relevant to recognize that, even if the process is described in terms of steps or parts, no implication is being made that the analysis process is linear or that one step must be before the next one is started.

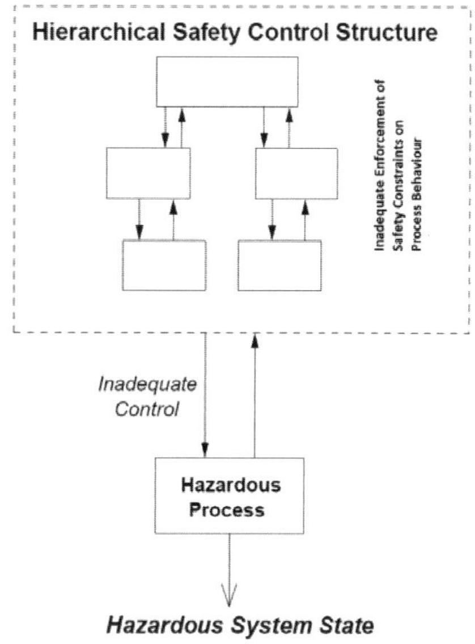

Figure 2: Relation between accident and inadequate enforcement of safety controls [7].

- Identify the system(s) and hazard(s) involved in the loss.
- Identify the system safety constraints and system requirements associated with that hazard.
- Document the safety control structure in place to control the hazard and enforce the safety constraints. The structure has to include the roles and responsibilities of each components in the structure itself as well as the controls provided or created to execute their responsibilities and the relevant feedback provided to them. This structure has to be completed in parallel with the later steps.
- Determine the proximate events leading to the loss.
- Analyze the loss at the physical level, identifying the contribution of each of the following to the events: physical and operational controls, physical failures, dysfunctional interactions, communication and coordination flaw. Determine why the physical controls in place were ineffective in preventing the hazard.
- Moving up the levels of the safety control structure, determine how and why each successive higher level allowed or contributed to the inadequate control at the current level. For each system safety constraint, it can happen that the responsibility for enforcing it was never assigned to a component in the safety control structure; in case it is assigned, it might happen that a component or components did not exercise adequate control to ensure their assigned responsibilities (safety constraints) were enforced in the components below them. Any human decisions or flawed control actions need to be understood in terms of (at least): the information available to the decision maker as well as any required

information that was not available, the behaviour-shaping mechanisms (i.e. the context and influences on the decision-making process), the value structures underlying the decision, and any flaws in the process models of those making decisions and why those flaws existed.

- Examine overall coordination and communication contributors to the loss.
- Determine the dynamics and changes in the system and the safety control structure relating to the loss and any weakening of the safety control structure over the time.
- Generate recommendations.

The final recommendations are the important conclusions to determine how to eventually modify the safety-control structure in the most cost-effective and practical way to prevent similar accident processes in the future.

The CAST approach has been extensively applied in several fields like defence [9], healthcare [10, 11, 12], aviation [13, 14], railway [15, 16] with remarkable results. Moreover very extensive source of information and interesting basis for discussion on its application are available on STAMP Workshops website (https://psas.scripts.mit.edu/home). The aim of this paper is to possibly extend its application to enhance ship safety by systematically analysing real accidents, as already mentioned in the introduction. Some very recent applications in the maritime fields already exist, both in the Navy [17] and in the merchant field [18, 19].

3. THE APPLICATION CASE

To assess the CAST suitability for the maritime field, an application to the Herald of Free Enterprise accident has been carried out. The selection of this specific case is motivated both by its worldwide impact and significance for the maritime community and the relevance for the intense safety rulemaking activity at IMO derived by that tragedy [20]. In fact, shortly after the accident the United Kingdom came to IMO with a request of a series of emergency measures to be considered for adoption, mainly as amendments to the International Convention for the Safety of Life at Sea (SOLAS).

Most of the proposals were based on the findings after the inquiry into the disaster [21], covering a very wide range of safety issues, and only few of them are mentioned in the following: indicators to be provided on the navigating bridge about the closure of all doors, new requirements for stability assessment concerning both the intact and the damaged condition, the introduction of SOLAS Chapter IX "Management for the Safe Operation of Ships" to make the International Safety Management (ISM) Code mandatory.

In relation with this last IMO initiative, it is worth mentioning that the accident investigation report put a strong accent on the relevance of operational and organizational issues in the evolution toward the disaster. Such aspects are issues that the CAST system-thinking approach is claimed to address appropriately, as described in the previous paragraph.

3.1 THE HERALD OF FREE ENTERPRISE ACCIDENT

The ship

The MS Herald of Free Enterprise was a roll-on and roll-off (RO/RO) ferry, built in 1980 for the company Townsend Thorensen. She was registered at the port of Dover and intended originally for operations between her registered port and Calais. However in the occasion of the accident the ferry was in service on the Dover – Zeebrugge route.

The propulsion was made of three diesel engines for a total continuous rating service power of 17880 kW, delivered on three shafts provided with controllable pitch propellers. The service speed was 22 kn.

The first vehicle deck above the water level (G deck), was a through vehicle deck and provided with a single weather tight door at stern and double weather tight doors at the bow.

RO/ RO ferry are a very successful ship typology from the commercial point of view, but undeniably the large flat vehicle deck located near the water level (convenient for loading and unloading purposes) is a typical characteristic that requires an high level of attention during the operational activity of the ship. Flooding of this area is hard to contain and it is well-known that when a large volume of fluid moves on this space, the stability of the ship becomes a critical issue.

Because of this aspect, further attention is given to the weather tight doors located fore and aft on the main deck (G deck). The doors are hydraulically operated, they split in the middle and open to either side, resting on the side of the ship when open. When closed, the doors are pressed together in the middle and a neoprene seal around the perimeter of the doors is activated. The doors were designed to withstand any forces during navigation and were designed to be as strong as the adjacent parts of the bow or stern hull.

Beside the physical description of ship, also the operational environment i.e. the harbour berth in Zeebrugge deserves attention. The port of Zeebrugge was not ideally compatible with the loading/unloading activities of the Herald, because of a high discrepancy between the port loading ramp and the vessel's highest loading deck, the E deck. This was particularly evident in case of unfavourable tide. In order to cope with it, the bow of the vessel was ballasted in order to better align the deck line with the ramp. This was obtained acting with the ballast tank system and it was a quite long process.

The formal investigation reported that on the 6th march 1987 the ship was significantly overloaded on that occasion and moreover draught marks have not been read. At the same time, because of shortage of time to conclude the de-ballasting operations, the ship left the Zebrugge harbour still trimmed by bow.

Other significant issues to better understand the accident will be delivered during the development of the CAST application case.

The evolution toward the accident

It was standard practice for the assistant Bosun to close the main deck doors before departure. On that day, the assistant Bosun had taken a short break after cleaning up the car deck upon arrival to Zeebrugge. When the departure alarm sounded he was asleep and did not hear the call. Anyhow also the First Officer was usually required to stay on the main deck until the doors were closed, but on the day of the disaster he had returned to the wheelhouse in an effort to speed up the departure of the ferry.

For the same reason, i.e. urgency for departure, the vessel left the berth before the bow ballast tanks had completely emptied. The ferry exited the harbour with all the three engines set to full speed and the ship accelerated to a speed of 18 knots. At that speed a combination of steady wave profile, ship static and dynamic trim, shallow water (squat) effect caused water to enter over G deck at bow. At a speed of 18 knots it is believed, through full-scale experiments with the sister ship Pride of Free Enterprise, the G deck at bow was approximately two meters below the water level. Water entered over the garage deck and in short time the ship capsized coming to rest on the starboard side on a sand bank at a heeling angle of approximately ninety degrees.

3.2 THA ACCIDENT ANALYSIS WITH CAST TOOL

As evidenced in the description of the methodology, some steps are to be fulfilled in order to develop the CAST application.

An accident is considered as an undesired event that results in a loss (including loss of human life, human injury, property damage, environmental pollution, …).

In this application case two accidents have been identified:

- the ship capsizing;
- the inability of rescuing all people after the capsizing.

This second issue has taken a large part of the formal investigation and it has been source of significant recommendations to IMO to improve ferry safety. The application of CAST can also take care of this very important part, but, purely for sake of concision, the paper will focus only on ship capsizing. Such matter is deemed to be enough significant to permit the discussion about CAST applicability in the maritime field.

Accordingly with what above, the systems and the hazards involved in the loss are identified together with the system safety constraints and system requirements associated with hazards.

An hazard is a system state or a set of conditions that, together with a particular set of worst-case environment conditions, will lead to an accident (loss).

Since in this case the capsizing of the ship is connected with the ingress of water on the garage deck, this last event is assumed as the significant hazard to take care of.

The ship physical systems specifically involved in the accident are the bow doors at G main deck and the ship overall residual freeboard at bow, in turn related with the actual loading condition and the employment of the forward ballast tanks for trim adjustment in harbour operations.

In Table 1, a recapture of the safety constraints, that should have prevented the flooding of G deck, is presented.

HAZARD	SYSTEM SAFETY CONSTRAINTS
Flooding on the Garage Main Deck	Closure of the bow doors before departure Reading of draft marks before departure Emptying of the ballast water tanks before departure

Table 1: Hazard identification and System Safety Constraints.

A representation of the safety control structure to control the hazard and enforce the safety constraints can be delivered by the diagram shown in Figure 3.

The significant physical systems for the accident analysis, already mentioned, are identified in the lower part of the diagram. Then, in the upper part, the safety control structure to enforce the safety constraints are described. This model could be even larger, entailing further control entities above the ship management company, such as Government, Regulatory Agencies, Industries Associations, User Associations, Unions, Insurance companied and so on. At the same time, it could be even more detailed both at the level of ship physical description and at the higher levels of the safety control structure. At present, the application is formulated with an essential initial model, where a fundamental role is also played by the relationships among the system components represented by the arrows. Basically, all the top-down arrows express a control activity and all the bottom-up arrows express a feedback activity.

They are numbered in order to be better described in Table 2.

As a further step forward, in the following the proximate chain of events are presented together with the recapture of the additional information.

Proximal chain of events

- The Herald of Free Enterprise was originally intended for the service on the Dover-Calais route, nevertheless she was operating between Dover and Zeebrugge.

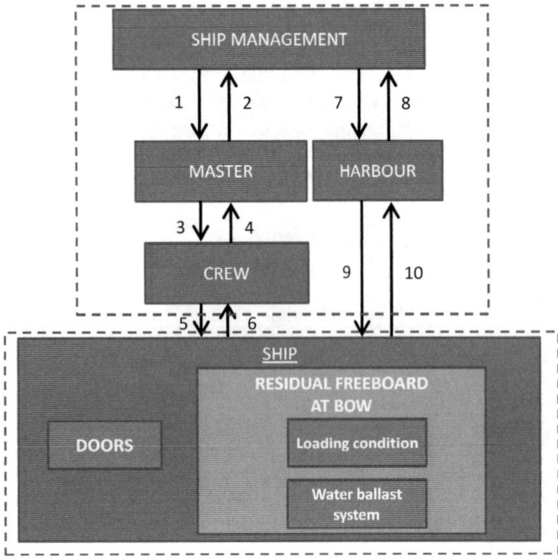

Figure 3: the safety control structure for the Herald of Free Enterprise accident in terms of CAST approach.

Communication Channels between levels		
CTRL	FDBCK	
1		Time scheduling, Payload, clear procedures, training, safety culture.
	2	Experiences, feedbacks, report about critical issues.
3		Instructions and procedures for operations (sailing and ship speed included).
	4	Feedbacks on operations (including execution/not execution of the actions).
5		Loading operations, departure operations, ballast management.
	6	Draft trim heel, tightness, speed, dynamic trim.
7		Selection of adequate harbor facilities for operations.
	8	Harbor characteristics and availability.
9		Procedures for single loading level, shallow water.
	10	Trim by bow, speed limits to avoid squat effect.

Table 2: Description of arrows of Figure 3.

- On the 6th of March 1987 afternoon, in the Zeebrugge harbour, activities for unloading, cleaning and loading the ship for the route back to Dover were underway. The assistant Bosun, after his activity went to his cabin, where he fell asleep.
- The loading activities on E vehicle deck implied the use of the ship ballast water system to obtain a ship trimmed by bow (necessary due to the specific characteristics of Zeebrugge berth).
- After loading activity completion of E vehicle deck, the loading activity proceeded then with the loading of vehicles on G vehicle deck (main deck) and the discharge of fore ballast water system started.
- Loading operation were completed with few minutes of delay on the time scheduling.
- The ship was overloaded, but apparently there was no awareness/evidence of this situation.
- The ballast water tank 14 was not completely empty yet, so the ship still had a trim by bow.
- The assistan Bosun, in charge of closing the bow doors, was not awakened by the call "Harbour Stations", which was given over the Tannoy address system. He didn't close the bow doors.
- The loading officer didn't check the doors closure.
- The Master didn't receive any information about bow doors and he assumed them closed.
- The Herald of Free Enterprise left berth 12 in the harbour of Zeebrugge at 18.05 (with few minutes of delays).
- The Master increased the ship speed from combinatory 4 to combinatory 6 (ship acceleration form 14 to 18 knots).
- The combination of static trim by bow, dynamic sinkage and trim, squat in combination with shallow waters and increase of the ship steady wave at bow caused a significant head of water above the G deck level at bow.
- Water entrance on G deck.
- Free surface effect on G deck.
- At time 6.28 pm progressive increment of heel angle and capsize (the ship touched a sand bank).

Additional information

Manning

- Operations in Zebrugge harbour were different in comparison with Calais (see next point), therefore the duties of the deck officers at Zeebrugge should have been defined accordingly. Moreover, three crews and five sets of officers were employed in manning the Herald. As a consequence, the officers did not always have the same crew. This made even more essential uniformity in the duties of each set of officers and of the members of each crew.

Zeebrugge harbour

- No. 12 berth at Zeebrugge was a single level berth, not capable of loading both E and G decks simultaneously, as at the berths in Dover and Calais for which the ship was designed. The ramp at Zeebrugge was designed for loading on to the bulkhead deck of single deck ferries. In order to load the upper E deck of the Herald, it was necessary to raise the ramp. When loading or unloading with unfavourable tides, the ramp could not be elevated sufficiently to reach E deck. As the ships berthed bows to this berth, it was necessary to trim the ship by bow, to allow the raised ramp to reach E deck. To achieve this trim, ballast tanks No. 14 and 3 were filled. No. 14 had a capacity of 268 m^3, and No. 3 a capacity of 42 m^3. The general practice was to start flooding No. 14 deep tank about two hours before arrival in order to berth with the requisite trim by bow. The ballast system could fill or empty No. 14 ballast tank at a rate of 115-120 tonnes per hour.

The loading condition

- The Herald was manned by a crew of 80 hands and on the 6th of March in Zeebrugge was laden with 81 cars, 47 freight vehicles and three other vehicles. Approximately 459 passengers had embarked for the voyage to Dover. The ship was overloaded. It was not possible to check actual ship draft on ship draft marks.

Special care when sailing with trim by bow

- The Master and deck officers testified that when entering or leaving Zeebrugge with this ship trimmed by bow, they reduced speed to a level which would avoid the water coming over the bow spade.

Pressure to leave the berth

- The sense of urgency to sail at the earliest possible moment was very strong among officers. To spare time, the order “Harbour Stations” was sometimes given before loading was complete, taking advantage of the few minutes necessary for the crew to arrive at their stations.

Previous near accidents

- Before this disaster there had been no less than five occasions when one of the Company’s ships had proceeded to sea with bow or stern doors open.

What above is relevant to complete the description for characterizing the evolution toward the accident.

The next CAST steps forward will strongly rely on the diagram of Figure 3 and on Tables 1 - 2 and on proximal chain of events coupled with additional information.

In fact, the next analysis of the loss at the physical level and the possible inadequate controls at the level of safety control structure imply to identify the contribution to the events of physical and operational controls, dysfunctional controls or feedbacks, communication and coordination flaws. In this perspective, the Herald of Free Enterprise accident is of outstanding interest because no physical failures can be identified and this is very helpful to clarify that safety is different from reliability [8].

In Table 3, a scheme of the main issues relevant to the loss at physical level is presented. As it can be observed, the focus of the analysis is about the physical systems as requested, however, the raised problems are mainly control issues. The strength of the CAST approach is to provide a framework where this kind of flaws can be spotted and put in evidence.

Then, the analysis is focussed on a higher level regarding the safety control structure and the target is to determine how and why each successive higher level allowed or contributed to the inadequate control of safety constraints at the current level.

It is well-recognized that this accident was not caused by the lack of reliability of one or more components, but by flaws in the safety control structure. This is the reason why the present step of the approach is the more significant. In fact, the accident was due to the absence of a well-defined control structure, especially in relation with the enforcement of safety constraints, i.e. to check the execution of the responsibilities of the officers and of all the other members of the crew.

ANALYSIS OF THE LOSS AT PHYSICAL LEVEL	
Safety Constraints Violated	The bow doors were open at ship departure; the ship was overloaded and trimmed by bow.
Failures and inadequate controls	No systems for the assessment/feedback of bow doors closure; no systems for the assessment/feedback of the actual ship draft (sinkage and trim).
Dysfunctional interactions	Pressure to speed up operations and departure; the increasing of the ship speed aroused the trim by bow and the relevant water head on G deck; the operational procedures in Zebrugge harbour were different than in Calais; the crew/officers staff had a frequent turn over.

Table 3: Recapture of analysis of the loss at physical level.

From the analysis of what happened, some different flaws at the different safety levels can be pointed out and they are ordered in the framework provided by Tables 4, 5, 6.

ANALYSIS AT SUPERIOR LEVEL OF THE CONTROL STRUCTURE – THE COMPANY	
Safety Related Responsibilities	Provide and maintain safety culture; provide rules and procedures for safe operations; provide specific training for the loading operations in Zeebrugge harbour; provide the ships with suitable equipment to support crew in frantic ship operations (indicators for the closure of the doors, reliable draught indicating system, more powerful ballast pumps).
Unsafe decisions and Control Actions	Insufficient control on correct enforcement of the operational procedures on board; insufficient response to feed-back and report from ship officers, especially in relation with the devices for assessing the ship drafts and the bow doors conditions; insufficient attention to the different features of the Zeebrugge harbour.
Dysfunctional interactions	Believe that the "self-management" of activities on board was the best procedure.

Table 4: Recapture of analysis at superior level of the control structure – the Company.

ANALYSIS AT SUPERIOR LEVEL OF THE CONTROL STRUCTURE – THE MASTER	
Safety Related Responsibilities	Follow and practice Administration and Company rules for safe operations; check the loading conditions; check the ballast system conditions; check the bow doors conditions.
Unsafe decisions and Control Actions	Left the harbour without having feedback about the status of the fore tight doors and the actual ship draft in terms of sinkage and trim.
Dysfunctional interactions	Pressure to speed up the operations and departure; the operational procedures in Zebrugge harbour were different than in Calais; moreover the crew/officers staff had a frequent turn over

Table 5: Recapture of analysis at superior level of the control structure – the Master.

ANALYSIS AT SUPERIOR LEVEL OF THE CONTROL STRUCTURE – THE CREW	
Safety Related Responsibilities	Follow the Master instructions, according to the Company instructions and Administration rules; close the bow doors; read the actual ship draft; management of the ballast water system; check the closure of bow doors and the draft conditions; inform the Master about the bow doors closure and the actual ship waterline.
Unsafe decisions and Control Actions	The assistant Bosun did not close the fore tight doors; the First Officer did not assess the closure of the fore tight doors and the draft conditions; general absence of communication to the Master about actual bow doors and waterline conditions.
Dysfunctional interactions	Pressure to speed up operations and departure; the operational procedures in Zebrugge harbour were different than in Calais; the crew/officers staff had a frequent turn over.

Table 6: Recapture of analysis at superior level of the control structure – the Crew.

It is now possible to came up with a general assessment about the organization and management of the ship, her Master and crew, but also at the level of the Company management. Actually, it seemed as all the ship operations run on an everyday experience-based way, relying on the general principle that the previous time it worked properly. However, this working practice could not safely face a different scenario from the usual one, as proven in this accident.

To this regard, it is interesting to comment how, in the past, five times sister ships of Herald of Free Enterprise left the harbour with bow doors open without any accident.

Making reference to the safety constraints evidenced in Table 1, it appears how an efficient control on the freeboard could have helped in preventing the loss also in this case. Such evidence supports the very useful large scale view provided by the CAST safety control structure, where many considerations can take root: in this framework it is possible to observe clearly how the banality of not asking/providing a feedback about the

execution of an action (closure of tight doors) caused a very critical hazard. This however, if happened alone, could even not evolve in a ship loss, as in fact experienced from other ships of the same class. However, such hazard together with a trim by bow and speed increase in shallow water crated the condition for very fast and unstoppable water flow on the garage deck. So a flaw in the operational management comes up: due to the hurry of leaving the harbour, the Master left with a trim by bow further increased by ship speed and shallow water effects. And from operational management the focus can shift now to the ship physical system: the pumps were undersized and so the ballast management was too long, being an operational constraint.
The high level nature of the Company "control/feedback loop" is evidenced in the safety control structure of Figure 3 and the most significant relevant flaws for this accident are recaptured in table 4: clear procedures, training, audit, communication between the ship and the shore office are among the most powerful resources to enforce the safety constraint of table 1 and are also the among the pillars of ISM Code, one of the most significant IMO countermeasure following the accident, that specifically focus on safety management.
At the same time, it can occur that, even with the presence of an hazard, nothing dangerous happens because ships are built with an intrinsic resilience, that nevertheless can be not sufficient to face unexpected scenarios and this issue should be addressed in a design phase. The sufficient level of resilience designed into the ship could be discussed as well by means of another STAMP system-based tool, named STPA (System-Theoretic Process Analysis) [22], to be applied this time during the design phase with a proactive approach instead of a reactive one.
After the fulfilment of all the CAST steps, general recommendations are presented in Table 7.
They can be developed on two different levels, i.e. in relation with the physical system and with the safety control structure.
It is worth mentioning that most of the issues identified in the general recommendations have been tackled by IMO rulemaking activity along the years immediately following the accident.
In this perspective the introduction of the International Safety Management (ISM) Code, among the important IMO initiatives, is the answer to the flaws of the safety control structure.

4. CONCLUSIONS

CAST has proved to be an excellent tool in the maritime field in order to document and analyse the results of an investigation process. Starting from the official investigation report on the Herald of Free Enterprise accident, thanks to the concepts of "safety control structure" and "safety constraints", it has been possible to put in evidence why the event occurred and to identify the changes that could prevent a similar accident in the future.

GENERAL RECOMMENDATIONS	
Physical system & operations	Provide indicators on the navigating bridge for all the doors which, if left open, could lead to major flooding of a ro-ro cargo space.
	Ships must have scales of draft marked clearly at the bow and stern. In case these are not easily readable the ship must also be fitted with a reliable draught indicating system. After loading and before departure, the Master must determine the actual ship's trim and stability.
	The ship must have an approval certificate for operating from a specific berth with particular attention to the possibility to shut their bow and stern doors without moving from the berth.
Safety control structure	Companies are required to prepare plans and instructions for key shipboard operations and to make preparations for dealing with any emergencies which might arise.
	Regular survey and audits should be held by the company to ensure that the safety rules are complied with and the system itself should be reviewed periodically to evaluate its efficiency.
	Implement a proper channel of communication between the ship and the shore office.

Table 7: Recapture of General Recommendations.

One of the most important added values of CAST is, in fact, the safety control structure model, where the main systems and controllers are identified in order to discuss what was wrong at the level of components, interactions and responsibilities. The introduction of different hierarchical levels and the control/feedback channels permit to discuss the influence of Human Factor, in a deeper perspective than the human error of the single person to blame, evidencing why the complex socio-technical circumstances has conditioned the human being behaviour.
The application, in fact, has been carried out not to add further conclusions beside the investigation report, but to implement a new analysis of the investigation report in terms of system thinking.
Actually, also in the present application the safety control structure could have been wider involving superior entities at level of control structure and a larger domain as far as the physical system is concerned.
In other fields of application, CAST has proved the capability to analyse accidents where a superior level of technology is involved, e.g. automation systems. In such case a more complex control structure can be modelled

as well, with the possibility to analyse the interactions between human control and automated control.
As a final comment about the importance to implement an integrated reactive and proactive approach for ship safety, STAMP methodology can provide tools for both these safety pillars: CAST, described in this paper, for the reactive part and STPA (System-Theoretic Process Analysis) for the proactive initiative that is rooted in the same system-thinking approach and can be used to guide the design process in order to design safety into the ship.

5. REFERENCES

1. SOLTANI, A., 'Proactive maritime safety: concepts and applications' *World Maritime University Dissertations 254,* 2009.
2. TARELKO. W., 'Origins of ship safety requirements formulated by International Maritime Organization' *Procedia Engineering 45,* 2012
3. PSARAFTIS, H.N., 'Maritime Safety: to Be or Not to Be Proactive', WMU *Journal of Maritime Affairs, No.1, 3-16*, 2002.
4. SCHRÖDER-HINRICHS, J-U., PRAETORIUS, G., GRAZIANO, A., KATARIA, A., BALDAUF, M. 'Introducing the Concept of Resilience into Maritime Safety', *Proceedings of 6th Symposium on Resilience Engineering: Managing resilience, learning to be adaptable and proactive in an unpredictable world,* Lisbon, Portugal, 2015.
5. SCHRÖDER-HINRICHS, J-U., HOLLNAGEL, E., BALDAUF, M., HOFMANN, S., KATARIA, A., 'Maritime human factors and IMO policy', *Maritime Policy & Management, 40:3, 243-260*, 2013
6. CARIDIS, P., 'CASMET - Casualty Analysis Methodology for Maritime Operations, Final Report for Publication' *NTUA Report*, 1999.
7. LEVESON, N., 'Engineering a Safer World: Systems Thinking Applied to Safety', *MIT Press*, 2011
8. LEVESON, N., 'Applying Systems Thinking to Analyze and Learn from Events', *Safety Science 49(1):55-64*, 2011.
9. LEVESON, N.G; ALLEN, P. & STOREY, M-A., 'The Analysis of a Friendly Fire Accident using a Systems Model of Accidents', *International Conference of the System Safety Society*, 2002.
10. DÜZGÜNA, H. S., LEVESON, N., 'Analysis of soma mine disaster using causal analysis based on systems theory (CAST)', *Safety Science*, 110 2018.
11. LEVESON, N., DAOUK, M., DULAC, N., & MARAIS, K., 'Applying STAMP in accident analysis', *NASA Conference*, 177–198 , 2015.
12. LEVESON, N., SAMOST, A., DEKKER, S., FINKELSTEIN, S., RAMAN, J., 'A Systems Approach to Analysing and Preventing Hospital Adverse Events', *Journal of Patient Safety*, 2016.
13. NELSON, P. S., 'A STAMP Analysis of the LEX Comair 5191 Accident', *Control*, 38, 2008.
14. THOMAS, J., MALMQUIST, S., 'Learning from Accidents that are a Consequence of Complex Systems' *The International Society of Air Safety Investigators Conference*, 2017.
15. OUYANG, M., HONG, L., YU, M., FEI, Q., 'STAMP-based analysis on the railway accident and accident spreading : Taking the China – Jiaoji railway accident for example', *Safety Science*, 48(5), 544–555, 2010.
16. FAN, Y., LI, Z., PEI, J., LI, H., SUN, J., 'Applying systems thinking approach to accident analysis in China: Case study of ''7.23'' Yong-Tai-Wen High-Speed train accident', *Safety Science*, 76 , 2015.
17. KIM, T., NAZIR, S., OVERGARD, K. I., 'A STAMP-based causal analysis of the Korean Sewol ferry accident', *Safety Science*, 2015.
18. KWON, Y., 'System Theoretic Safety Analysis of the Sewol-Ho Ferry Accident in South Korea', *Thesis – Master of Science degree in Engineering and Management, Massachusetts Institute of Technology,* 2016.
19. MACKOVJAK, J.M., 'Systems Theoretic Accident Analysis of an Offshore Supply Vessel Collision' *Thesis – Master of Science degree in Technology and Policy, Massachusetts Institute of Technology,* 2016.
20. International Maritime Organization, 'IMO and ro-ro safety' *Focus on IMO*, 1997
21. Department of Transportation, 'The Merchant Shipping Act 1894, mv Herald of Free Enterprise', *Report of Court No. 8074 Formal Investigation, Crown Copyright*, 1987.
22. LEVESON, N., THOMAS, J.P., 'Handbook STPA' *copyright © 2018 by nancy leveson and john thomas,* 2018.

6. AUTHORS BIOGRAPHY

Paola Gualeni is graduated in Naval Architecture and Marine Engineering at the University of Genoa where she holds the position of Associate Professor. She teaches Ship Hydrostatics and Stability and her main fields of interest are Ship Safety and Innovative Ship Design Methodologies.

Costantino Bongermino in 2017 graduated in Naval Architecture and Marine Engineering at the University of Genoa where at present he holds the position of Ph. D fellow with specific interest on human factor integration in maritime systems safety, from both the design and the operations perspective. He serves as Sub-Lieutenant in Italian Navy on the Italian aircraft Cavour.

PRESENTATION

ADOPTING NEW TECHNOLOGIES – REDUCING RISK OR INTRODUCING NEW THREATS

R V Pomeroy, Southampton Marine and Maritime Institute, University of Southampton, UK

SUMMARY

The global demand for maritime transport continues to evolve as the new world order is established. At the same time the maritime industry is introducing new technology at an increasingly rapid pace, hastened by challenges to reduce costs and improve environmental performance. The likelihood is that future changes will include disruptive solutions driven by new technologies.

The introduction of novel solutions can deliver significant benefits but also bring unexpected outcomes. The paper considers the evidence from past introduction of new technology and the associated risks. The paper considers whether adoption of novel technologies could introduce new threats that will place unrealistic demands on the people in the system, recognising that the human element in also a major contributor to risk mitigation.

The issues discussed relate to the human element at organisation and individual levels, encompassing the whole life cycle - concept, design, construction, maintenance and operation.

1. INTRODUCTION

The safe and efficient transport of people and goods by sea has a long history of cautious evolution, with ship owners adopting new methods from time to time as the trading patterns changed and new technologies became available. This introduction of novelty was once conducted in a gradual manner, gaining experience and refining design prior to widespread use. The idea of an 'experimental' class notation was accepted as a sensible approach, with the Rules of the classification society being developed on the basis of initial service experience. Regulation was heavily influenced by serious incidents and was necessarily responsive rather than anticipatory with regard to technology change.

Now the industry is faced with much more rapid change, due to an incessant drive for more efficient transport and the introduction of new regulatory demands. There are now suggestions that change will also be driven by the availability of new technology, pushed into the market by suppliers rather than pulled by the user community.

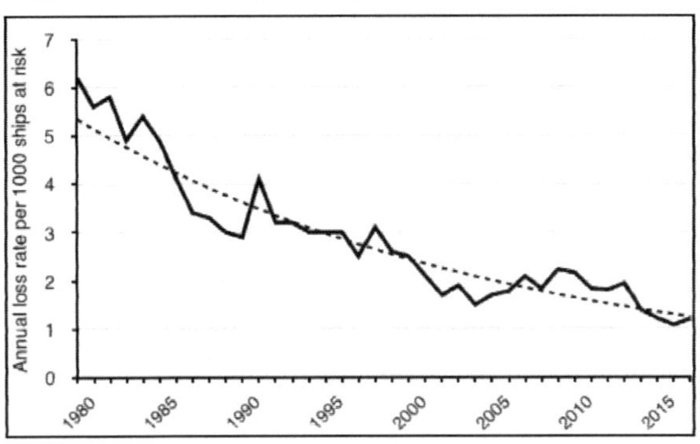

Figure 1: Annual total loss rate 1980 – 2017 [1]

In the recent past the industry has seen a progressive decline in major casualties. Figure 1 has been derived from a consistent source - the evidence base is addressed more fully in Section 4 of this paper with specific reference to the changes in technology that have occurred since 2000.

Perrow [2] introduced the idea that no matter how much effort was expended trying to reduce failure rates in complex technological systems there would always be some accidents, and he referred to these as 'normal accidents'. He also clearly considered merchant shipping to be a high-risk technological system because of the complex interactions between a web, both ashore and afloat, of involved parties, the natural environment and technologies that combine to create an error-inducing system which seems to defy efforts to improve safety performance. The inference is that the level of accidents will never be zero and that the human element in its widest sense is key to improvement.

Before leaving this illustration of safety performance it is useful to refer to the work of Duffey and Saull [3] who developed a model based on the idea that industries learn, from experience, and as a consequence the safety performance improves. The analysis of maritime casualty rates over a long period of time fits with the Duffey Saull Model. Their hypothesis is that the model asymptotes to a non-zero value, as indicated by the dotted curve in Figure 1, because the limiting factor is human activity and learning can never be perfect.

And yet, the maritime industry is faced with criticism whenever ships are lost, particularly when pollution results. With an apparent increase, for example, in the number of fires on container vessels and collisions, often involving well-found ships operated by established companies, there will continue to be calls for improvement from governments and insurers. The data presented in Figure 1 shows that the impressive reduction in the rate of total losses between 1980 and 2005 appears to have stalled, so can adopting new technology help, or might it hinder further progress?

This paper looks at the evidence from the incident record and considers how the adoption of new technologies could bring benefits in terms of improving safety performance, whilst cautioning against the risks that might accompany such change.

2. OPPORTUNITIES AND NECESSITIES

The pressure for adoption of new technology comes from several directions. The biggest driver for change comes from societal pressures linked to the great global challenges facing the world, including climate change, unequal distribution of wealth and the impact of population growth. Some of these challenges lead to the need to undo past successes as the evidence base changes. Shipping faces such issues, in particular the need to reduce dependence on fossil fuels which have been the basis of an efficient and reliable maritime transport system since the advent of steam and diesel propulsion. New options for energy sources, including batteries, alternative fuels such as LNG or methanol and wind, bring a new set of solutions that are unfamiliar and probably accompanied by unforeseen hazards. The energy challenge presents exciting opportunities for doing things differently, which could make a substantial change in the business of transporting goods and people. The change in ship design could be as radical as that seen when steam replaced sail, affecting the whole of ship design and operation, and all that implies for the people in the system. The change will also impact on the system beyond the ship with new facilities for fuel handling, all involving new designs and new working arrangements.

The industry has faced and is facing a number of regulatory requirements that results in the fitting of new technologies on ships, including the mandatory fitting of Automatic Identification Systems (AIS), Electronic Chart Display and Information Systems (ECDIS) and Ballast Water Management Systems. These requirements all bring to the ship new technologies that have not been fitted previously, accompanied by additional training and procedural demands. The introduction of a new solution provides a good opportunity to study how the process of change is managed and how the new technology is used in service. Bailey et al [4] studied the adoption of AIS over a period of four years following universal introduction in 2004. The study showed that learning during the early period of operation was slow but evident improvement in use was observed over the period, which resulted in greater confidence that the anticipated benefit would pull through. Perhaps more significantly in a discussion about new technology was the identification of practices that were not intended, or anticipated. These included the potential safety threat of an increased willingness to make ship-to-ship VHF calls to negotiate passing situations. The study also demonstrated that the new mandatory equipment was used for purposes that went far beyond the original intention of providing a means of identifying ships at sea. So much so that the original intention is rarely mentioned.

There is no reason why the findings of Bailey et al should be unique to the adoption of AIS, and it seems reasonable to assume that the generation of seafarers and ship managers that has been brought up with ready access to sophisticated electronic tools, including GPS-enabled hand-held navigation, will find for themselves uses of technology that were never envisaged by designers or those who elected to install it onboard ships. There appears to be a strange, slightly contrary, conflict between innate trust of digital technology and the willingness to subvert its intended use.

There has always been a keen interest in adopting new technical solutions to improve safety, particularly in an attempt to reduce human error. The confidence, although balanced against a warning of concerns of overreliance, is illustrated by the following comment by a major insurer:

"Technology could bring huge advantages for the maritime sector, catching issues early, before they escalate into a major casualty. Human error remains the biggest problem for shipping casualties but technology offers the potential to reduce human error, as well as reduce machinery breakdowns" [5]

Later in this paper the impact of automation on stoppage rates is discussed but the apparent strength of trust in technology, and in this context presumably increased automation through digital systems, is interesting. By the following year the same insurer was taking a slightly less optimistic view of the potential of autonomy to reduce human error when the following quotation was published:

"The majority of maritime accidents involve an element of human error, and autonomous ships could bring about a reduction in the frequency of incidents. However, autonomous ships will not remove human error completely. Human risk will still be present in the algorithms that drive the decision-making of autonomous vessels, while vessels continue to be monitored and controlled from manned onshore bases." [6]

Neither observation gives credence to the ability of humans to step in to automated systems and take actions that stop the escalation of incidents, apparently regarding the human as a problem rather than a solution.

The maritime industry is facing a period of rapid change, reacting to regulations and commercial necessities. In addition, the general advance in technology is providing opportunities to do things differently, with the most obvious being the possibility of developing unmanned ships. Any major change will only deliver a successful outcome when the balance between benefits and risks makes commercial sense, and that balance can readily be disrupted by the interaction with people. That means that in this time of rapid change understanding the interaction between people and technology throughout the whole chain from concept through design and manufacture and operation and maintenance, up to disposal is absolutely essential. The evidence, discussed later, suggests that the balance has not yet achieved.

3. DRIFT INTO DIGITAL DEPENDENCE

The pace of change is best illustrated by the advance of digital systems, which is now taken for granted. As the new millennium approached the principal focus of the IT specialists was on the large financial and data handling systems that were vulnerable to date handling constraints as the year 2000 arrived. It is quite commonplace now to hear the Millennium Bug, or the Y2K challenge, used to illustrate a threat that was overplayed by the IT sector that came to nothing, an example of crying wolf and a scare story that was exaggerated. In reality a lot of excellent people put in a great deal of effort to make sure that the known problems were dealt with in advance, but that was largely invisible and has rapidly been forgotten.

Many people never realised that a large number of electronic control systems were also found to be vulnerable, particularly where real time clocks were employed on circuit boards. With the focus on data handling systems it was a challenge to draw attention to a potential problem of unknown severity. For some time the author carried around a controller board taken from a simple supermarket freezer cabinet that included two real time clocks that only controlled a switching operation, but a real time clock was chosen because it was cheap and readily available despite there being no need for such a device. These were very simple devices with embedded code and both would fail on the Millennium switch.

Some shipowners would rely on information provided by suppliers of equipment whilst others conducted a major review of their inventories. These investigations found a number of unlikely equipments that could fail, often in an unpredictable manner. These vulnerable items of equipment included large winches fitted to offshore support vessels. It was, however, practicable to advise operators to revert to fully manual operation during the critical dates and times when systems were most vulnerable (7). Within a very short period of time that advice would no longer have made any sense as the rapid deployment of digital control technology affected most aspects of ships' machinery and equipment. This is most evident by the rapid uptake of the common-rail diesel engine with no camshaft. The digital electronics allowed greater flexibility in valve and fuel injection timing, which gave improved fuel consumption and greater control of exhaust emissions. However, the engine could no longer be operated without the digital electronics.

Of course, modern digital electronic systems, with ever increasing computing power and growing use of some form of intelligence, have permitted the integration of disparate items of machinery and equipment. This integration enables optimisation under a wide variety of operating conditions and encourages greater efficiency, which ultimately benefits society at large as the costs of transporting goods and people are driven down. The result is a complex system-of-systems, where far too often the integration is brittle because of design weaknesses. How complex has this digital revolution made modern shipping? Ships, other than cruise ships are generally perceived to be simple compared with, for example, an airliner but the reality is quite different. A modern cargo ship will carry, in items such as diesel engines, control room suites and bridge equipment, a similar volume of safety-related software, measured in terms of source lines of code, as a modern airliner – the question is whether or not it has been created and tested to the same level, or at least to the appropriate level, of integrity.

This increasing use of digital systems has crept up on the industry where the majority of seafarers and shore managers do not have more than a superficial knowledge of the underlying technology. Even amongst those who have been brought up in a digitally-connected world the knowledge is fundamentally about use rather than the technology, its limitations and the necessary integrity.

4. LEARNING FROM INCIDENTS

The research carried out by Duffey and Saull suggests that learning from accumulated experience is a key part of improving safety for any industry. When faced with new technology they showed conclusively [3] that the initial accident rate increases even where the technology is being introduced with the aim or improving safety and reducing human error. Their study was based on the introduction of 'modern glass cockpits', comparing the before and after safety performance on a single aircraft type for one major operator. That conclusion may be surprising when considering an industry that has a well developed training regime and support arrangements in terms of manuals at component and systems levels.

In a recent paper Pomeroy and Earthy [8] concluded that learning from incidents may not be appropriate during periods when technology changes rapidly. They noted that the maritime industry has become wedded to taking the lessons from history and embedding these into the safety regulatory framework. This approach has become so strongly instilled in the 'way we do things' that when new safety requirements are proposed without evidence that the undesirable outcome has actually occurred these are rejected as 'unjustifiable'. This conundrum will be revisited later.

4.1 CASUALTIES

Long term casualty data, as shown in Figure 1, can only give an indication of the overall safety performance unless the data can be disaggregated sufficiently so that the contribution of distinct causal factors can be determined. The downward trend in terms of annual total loss rate is very comforting but it is difficult to identify the effects of any particular action. The data includes both actual and constructive total losses and the determination of whether a badly damaged ship can be repaired economically depends on many factors, including the cost of steel, the price of scrap and the market situation.

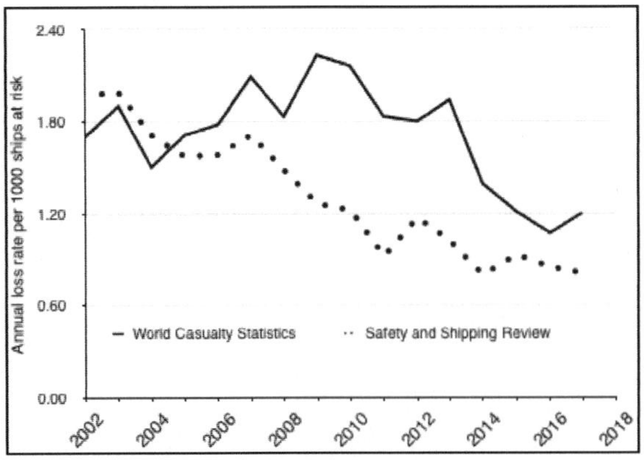

Figure 2: Annual loss rate 2002-2017

Figure 2, showing data extracted from the World Casualty Statistics [1] and from the various editions of the Safety and Shipping Review [9], shows the impact of adding back in constructive total losses that are late reported. It is clear that these two data sets suggest significant differences in the safety performance over the period 2002 to 2017. The author believes that the data extracted from the World Casualty Statistics is a more complete and correct representation of the situation, with the proviso that the values for the most recent two or three years will certainly see a significant uplift as additional losses are included following late reporting. The long-term improvement in the annual rate of total losses appears to have halted, which given the focus on safety improvement is disappointing.

When considering the role of the human element in maritime casualties, and incidents, it is useful to look in more detail at the categories of losses. Whilst the general consensus is that around 80% of casualties are the result of human error this figure is difficult to justify from the available data. Furthermore the definition of the human element, as declared by the International Maritime Organisation (IMO) [10], goes much further than the seafarer and following that definition there are very few incidents that do not involve some form of human activity within the causal factors identified through incident investigation.

Looking firstly at the causes of ship losses, Figure 3 has been derived from the data provided by Allianz [9] based on the Lloyd's List Intelligence Casualty Statistics for the period of sixteen years, 2002-2017. The casualty record also shows that losses peak during the winter months, in January in the northern hemisphere and in July in the southern hemisphere, suggesting a strong correlation with weather conditions.

Foundering is still the major cause of loss, accounting for 46% of total losses during this period. Recognising that the causes of some ships lost to foundering are not established conclusively, cargo movement features significantly in investigation reports,. Recent years have seen a number of bulk carriers lost (for example *Harita Bauxite*, *Trans Summer* and *Bulk Jupiter*) following liquefaction of dry cargoes. Ro-Ros are also vulnerable to shifting of improperly stowed cargo, as demonstrated by the case of *Hoegh Osaka* on departure from Southampton. Whilst the ship's crew has some responsibilities for the loading and stowage of cargo they rely on the work of others, so a complex human element issue. The aviation industry has identified ground handling, and particularly the loading of aircraft as a major safety concern, although the aircraft captain signs off the loading sheet, and there are some similarities with cargo shipping in terms of responsibility and control.

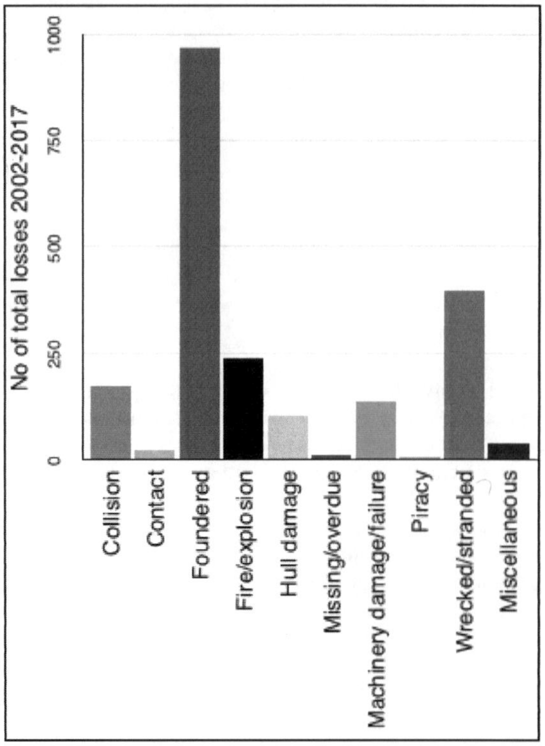

Figure 3: Causes of ship losses, 2002-2017 [9]

Navigation failures contribute significantly to the number of collisions (8.3%) and wrecked/stranded losses (18.9%) and those losses are showing some direct links to the new technology that has been added to the bridge in recent years. The only conclusion that can be drawn is that watchkeeping and navigating officers place a very high level of trust in the electronic aids to navigation without perceiving the need to carry out checks from time to time. This is not a new phenomenon, as the incident record demonstrates, for example:

- The cruise ship *Royal Majesty* ran aground in 1995 after the signal from the GPS antenna was lost and the deviation from the intended course was not picked up by cross-checking [11].

- The container vessel *Rickmers Dubai* was involved in a collision with a crane barge that was under tow, in 2014 [12]. The watchkeeping officer relied on the AIS content on the ECDIS display to detect other vessels in the area, and failed to notice the tug/barge combination that was crossing his track, and not carrying AIS.
- The cruise ship *L'Austral* ran aground in 2017 when the bridge team, and the pilot, relied on the electronic navigation systems for the night passage and lost awareness of their surroundings [13].

These examples all came with recommendations from the investigators relating to training on the systems in use, and on the general principles of navigation using modern integrated bridge systems with appropriate cross-checking and monitoring.

Fires and explosions have always featured in ship loss statistics but the change in recent years has been the prevalence of high impact fires on large container ships, without an external initiating cause such as a collision. These fires have proved difficult to contain, without injection of a major firefighting capacity, and have burned for long periods. The fire on *Maersk Honam* began on 6th March 2018 and the ship was finally permitted to berth on 27th July. The fire caused the death of five members of the crew and the ship was extensively damaged. As the level of manning on large ships has been reduced the ability deal with abnormal, but not unforeseeable, incidents is challenged. At the same time as ship size and cargo capacity has grown the risk has become much larger, without any significant change in the technical systems available to the crew to deal with fires.

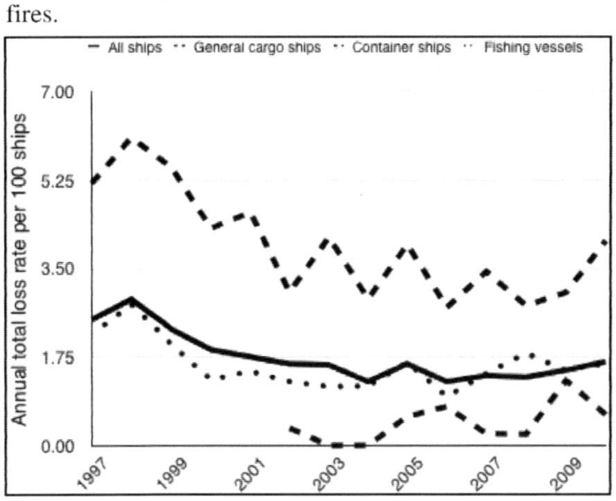

Figure 4: Annual loss rate for different ship types and all ships, 1997-2010

The other variable that is worthy of comment is the type of ships that contribute to the annual total losses. The IMO has for many years been considering the loss rate of general cargo ships and the data [9] shows that this ship type accounted for over 42% of all total losses over the period 2012-2017. The annual loss rate for general cargo ships, container ship and fishing vessels [1], for the period 1997-2010, is shown in Figure 4. The annual loss rate per 1000 ships at risk for general cargo ships is significantly higher than that for the whole fleet, the upper dashed line. The lower dashed line plot shows the loss rate for container ships. Perhaps more surprising is the dotted line which plots the loss rate for fishing vessels, but this excludes ships of less than 100 gross tons and is probably incomplete.

At this level the casualty data can only give an indication of the safety achievement over a period of time but it lacks detail. Even when broken down by ship types and cause the data lacks detail, but it is safe to conclude that the safety initiatives, including the major steps taken with the implementation of the ISM Code, the Maritime Labour Convention and major regulatory changes adopted by IMO, have not resulted in a continuation of the continual improvement in the annual loss rate evidenced in the period 1980 to 2000, or perhaps a few years later. The challenging question is whether this is because the safety initiatives were either ineffective or not implemented sufficiently rigorously or whether some other factor are acting against the safety improvements, presumably changes in technology, operation or business choice.

4.2 CHANGE OR NO CHANGE?

Although machinery failure is a relatively modest contributor to total losses, about 6.5%, reports from underwriters emphasise the importance of machinery damage in terms of insurance claims. The number of claims and the magnitude of individual claims are both increasing, despite the efforts to improve reliability and availability, now reportedly accounting for more than 60% of all hull and machinery claims.

The machinery installation has probably seen more change than any other system on the ship, and much has gone almost unnoticed. Not only has the use of automation and remote operation increased but so has the integration of machinery into more complex systems, even systems-of-systems. Alongside the changes to the installation itself there have been imposed changes to its operation. Running at slow speeds has become commonplace. Requirements for less polluting emissions has reduced fuel sulphur content. Rates of wear in diesel engines have increased as a consequence and more stoppages related to fuel have been reported.

Data relating to machinery reliability has always been notoriously difficult to find. However, an initiative by the IMO in 1981 [14] resulted in the presentation at subsequent meetings of the Sub-Committee on Ship Design and Equipment of the findings of various initiatives. One of the most useful papers, submitted by Poland [15], has been used in Table 1. The author presented this data [16], along with that from a study carried out by Lloyd's Register, which analysed data held within an early computerised planned maintenance system, and this is included in Table 1. The last column presents data for a modern fleet of ships, extracted from reports from ships. The incidents included are those that resulted in unplanned maintenance activities, for whatever duration. The surprising conclusion is that there is no change over forty years. The frequency at which ships experience unplanned maintenance activity remains at the same level. The significant difference is that for the modern fleet more than 30% of the incidents result directly from the failure of sensors. This situation requires more analysis to understand what is the reason for this lack of apparent improvement but an initial assessment does suggest that safety-related items, such as sensors, display a failure rate in maritime operating conditions that is too high. This is a chilling conclusion when the route to more automation and ships that are either autonomous or remotely-operated is to use a lot more sensors.

Component part affected	Polish study for IMO	LR study of PM data	Current fleet
Static parts	15.0	3.3	
Moving parts	23.9	33.7	10.8
Timing and gears	0.1	0.2	
Inlet and exhaust	3.2	25.7	22.6
Lubrication system	0.8	8.6	
Fuel system	46.6		5.5
Cooling system	1.7		
Control system	10.7		5.7
Starting system	0.9		6.2
Valves	-		15.7
No of defects per year	**11.4**	**17.1**	**15.5**

Table 1: Main engine stoppage rates and causes

Whilst machinery failures, and indeed hull structural failures, do still occur from time to time these almost invariably involve the human element amongst the causal factors although that may not be found amongst the crew. Failures do stem from undetected manufacturing and installation errors, misunderstandings about the operational requirement, poor design choices, and excessive trust in calculation tools. Reviewing recent reports of the investigations to establish the causes of incidents, there does appear to be an increase in the number of machinery failures that follow maintenance or repair where people have taken shortcuts or worked around the intended methodology to save time, inconvenience and cost. The removal and reinstallation of a propulsion diesel engine on the ferry *Wight Sky* [17] falls into this category where the hazard of introducing debris during reassembly was the consequence of electing not to remove the engine in one piece through a soft patch in the deck. This is the sort of incident that would not usually be regarded as due to human error, but clearly was the result of decisions that increased the likelihood of damage.

It is proving difficult to establish a robust evidence baseline that allows reliability trends to be compared. Where data is routinely collected by operators it is inevitably acquired for a different purpose, and so the taxonomy employed is rarely suitable. In many cases the cause is not established, beyond the most obvious conclusions, and this makes comparisons challenging. The biggest challenge is to establish where the friction occurs between people and hardware that leads to mistakes and errors, and failures. On the face of it there is a lot to be gained from collecting evidence, and using it in collaboration to mutual benefit.

4.3 MISUSE OR REALITY?

Reading through the reports of the investigations carried out to establish the probable cause of maritime incidents sometimes raises the question: *"was the equipment fitted to the ship being deliberately misused or was the behaviour displayed simply a facet of reality?."* It is often very difficult to tell, in retrospect, but there is a feeling that there is far too much trust placed in systems that do make life easier, at least at a superficial level.

The incidents involving the grounding of *L'Austral* [13] and the collision of *Rickmers Dubai* [12] with the crane barge seem to be illustrations of some of the unintended consequences of new technology on the bridge. It is worth recalling that Duffey and Saull found clear evidence that a change of technology could, or would inevitably in their model, lead to an increase in the incident rate. In the case of the change from conventional instruments on the flight deck to the modern screen layout there was also a change from direct readings to information that involved more hidden processing before display. The user no longer has to do the processing from the basic data but maybe the presentation of the result can give unwarranted confidence, and reduce the desire to check. That separation between the human and the data can lead to a loss of situational awareness and there is no simple solution. The impression given to the operator is that the displayed value is actually the data, and not a processed variable.

The watchkeeper on the bridge on *Rickmers Dubai* was content to trust the display on the ECDIS, showing AIS data, without reference to other navigation equipment. The fact that some obstructions might not show an AIS signal did not raise concerns. In the case of the grounding of *L'Austral* the blind navigation potential of the ECDIS was trusted, but when the crew noticed that the ship was off track the pilot was not challenged.

Designers need to be aware of the degree of trust placed in systems and to build in sufficient safeguards The reality is that the generation brought up in a digital world, and many who have become converts, do trust their gadgets, as witnessed by the number of people who blindly follow the instructions of their in-car satnav until they end up in the middle of fields.

5. REGULATION – BARRIER TO CHANGE OR FRIEND?

The commercial maritime sector uses a sophisticated, and complex, safety regulatory regime where two distinct strands are mutually entwined. The Rules of Recognised Organisations, essentially Classification Societies, and the Regulations contained within the International Conventions adopted by IMO and enacted through the national legislation of the National Administrations, the Flag States, each contain statements requiring compliance with the other. The Rules and Regulations have traditionally been heavily based on service experience, with a strong reactive implication. This approach is both a strength and, particularly when faced with rapidly changing technology, a weakness. The industry, whilst seeming to embrace a forward-looking proactive approach to regulation is still locked into seeking justification from experience in any new initiative [8]. It has been argued that this attitude is, itself, a barrier to adoption of new solutions that are safe.

Where collaborative research programmes have been undertaken, often funded by organisations such as the European Union, it has been common for the partners to include national administrations and/or classification societies who take on the role of developing a suitable set of regulatory requirements for the technology under investigation. Whilst this approach has the benefit of developing the supporting requirements in parallel with the technical solution the operational experience base is usually modest, limited to a few applications or even to laboratory demonstration. Nevertheless, the requirements will be based on sound philosophy that is founded on lessons from the past, including those relating to how technology is used. Van der Vleuten et al [18] make the compelling case that to engineer the future, and by inference to ensure that the level of attained safety is not compromised, we cannot afford to disregard the experience of the past. That seems to be forgotten when the technology is totally different and the proposers are not familiar with the underlying logic that is embedded in the existing requirements.

Some developers of new technical solutions, and particularly solutions that are disruptive and so clearly outwith the expectation of the originators of the requirements, suggest that regulation is a barrier to adoption. On the other hand the buyers of ships often regard the existence of classification society Rules or IMO regulations as an essential indication that the technology is sufficiently mature, and that the existence of standards somehow de-risks the solution. What is always missing is the human contribution, and experience has demonstrated that surprises will almost certainly come to pass in operation when people come to design, manufacture, operate and maintain the system.

One of the most helpful elements of regulation at the point of entry into service for a new technology is that relating to demonstration that the intended purpose can be met. The criteria for assessing performance and the methods for testing to confirm that the item performs satisfactorily are crucial for all concerned. Performance based standards, that are technology-agnostic and which are founded on demonstrating that essential functional safety requirements have been achieved, represent an effective approach when technology is changing.

In considering the limitations of learning from the past it was suggested [8] that some form of synthetic testing could be used to provide a substitute for operational experience. This would obviously need to be both rigorous and of sufficient duration to provide any meaningful advantage but could prove to be a route to the creation of effective safety regulation that would remove perceived barriers to deployment without any compromise in maritime safety. It would be a challenge to develop a robust simulation route, and in discussion Duffey [19] questioned whether a synthetic approach would ever be as effective as encapsulating real experience, but it would be preferable to relying on regulations that did not embrace human behaviours.

Regulation of an international industry brings some major advantages, along with challenges about universal enforcement, and it also underpins the procurement and maintenance practices for shipping. The benefits should not be lost, but where new technical solutions are adopted an effective route to demonstrate the attainment of at least an equivalent level safety is required. Whilst the IMO has introduced guidance on the demonstration of equivalence to alternative arrangements in substitution of those prescribed by the Safety at Life at Sea Convention [20] based on a risk assessment process, this begs the question whether the risks associated with the benchmark are understood.

In a time of change Regulation can become a barrier to adoption if the necessary steps are not taken.

6. GROWTH OF AUTOMATION

The shipping industry faces the introduction of novel technology for several applications, not least for the treatment of ballast water and emissions to atmosphere from diesel engines. Bridge equipment will continue to evolve. All will involve the human element, throughout the product life-cycle, and have the potential for adding to the burden on people engaged in the operation of the asset and possibly for creating an as yet unimagined pathway to a maritime incident.

However, the strand which continues along a pathway towards possibly unmanned and even fully autonomous shipping is the continual growth of automation, and the increasing dependence on digital systems. Along that pathway the marine engineering installations have become increasingly complex and integrated, and the people have become shielded from the machines by more and more layers of 'processing' as data is converted into information.

6.1 EARLY DAYS OF AUTOMATION

Automation in ships was first raised within the Inter-Governmental Maritime Consultative Organisation, later IMO, in 1963 with a note from the Secretariat presented to the Maritime Safety Committee in 1964 [21]. This short paper defined automation as *"those processes in which machines – often including electronic controls – adjust and control their own performance with little or no human intervention, once the operation is started"*. The definitions offered for fully-automated and partly-automated systems, and indeed for remote control, are still appropriate, although the technologies employed in automation have changed dramatically in the intervening years. The Committee was invited to consider the definitions offered, the effects of automation on manning and safeguards to maintain safety at a level equivalent to the then current level. Interestingly the paper also mentions the social implications, if only to exclude these from discussion. No mention is made specifically about the relationship between people and the automated system, although when considering a partly-automated system the paper notes that *"the human element* (author's comment - is this the first use of this terminology?) *may consist of overall control, supervision of operation and manual resetting if errors* (author's comment – errors presumably by the automation)*..."* which presents an interesting perspective.

The early steps into automation in the British and Commonwealth Shipping Company (Union Castle and Clan Lines) were presented by Munton et al in two papers [22, 23], describing the introduction of a new generation of motor ships, starting with *Clan Macgillivray* class, which featured operation of machinery from a control room. In the earlier paper they began with *"Obviously there are some confused and conflicting opinions and ideas on the subject of automation, ranging from those (generally non-technical) who think it is the answer to all problems and, at the other end of the scale, those (generally engineers with a sad experience of early automatic controls) who want nothing to do with the subject"*. The same could be written today around autonomous ships.

The first paper [22] describes in detail the progress with steam ships and the diesel ships in instrumentation and control technologies, from pneumatic and hydraulic devices to electrical and electronic. In looking to the future the authors stated *"..that in the immediate future, development should be concentrated on the operation of the machinery from a control room within the machinery space.*
The next step, and it will be a major step, will be to take the control room right outside the machinery space, with the engineer on watch making only an infrequent – possibly once per watch – tour of the machinery space".

They then refer to the loss of direct sensory experiences when the engineers are comfortably housed in an air conditioned and sound-proofed control room, noting the importance of touch and smell which are not possible to replicate artificially. However, they end up by wondering if it might be possible to eventually transfer the control room to the bridge, without loss of safety.

In their second paper Munton et al [23] stated that the development had *"the limited aim of reducing the number of watchkeepers in the machinery space"*. They presented data on the number of watches where alarms triggered some action, which represented 7-10% on the three ships of the *Clan Macgillivray* class. The authors also described the next series of ships, concluding that *"there can only be two reasons for extending the use of automation, humanitarian and economic"*. This represents a clear statement of purpose – to save money and to provide a better working environment for the crew. With hindsight, it is interesting to read how quickly the benefit of reducing crew costs was taken for an early adoption of widespread new technology.

The two papers and the published discussion are fascinating. There is a thread, noting factors such as increasing costs of crewing, that pursues the traditional argument for automation – it will replace the need for human control. There is also a parallel undercurrent that notes that people will still be involved in operation of the ship's machinery, but they may not be patrolling the machinery spaces. The tasks of the humans will be different, requiring different knowledge and skills.

This historical resource gives a detailed account of the installations, the technical approaches and the pace of development (even within the three years spanned by the two papers) but it totally supports the 'irony of automation' noted by Bainbridge [24] that *"the more advanced a control system is, so the more crucial may be the contribution of the human operator"*. In the record of maritime incidents there are several cases when faced with a situation where an automated action has occurred the crew has failed to understand what is going on, and has made a decision that has resulted in a minor event escalating to a major incident. The contact between *P&OSL Aquitaine* and the berth [25] illustrates how people can fail to unravel a complex situation when the automation acts and they are called into a part-developed incident pathway, with little time to react.

These papers were written over fifty years ago and yet the questions posed by the authors remain valid although the technology of automation has advanced considerably. The journey has passed from simply taking signals from machinery to instruments within a control room, through the use of display screens to integrated control desks, as illustrated in Figure 5. This parallels the development within other industries, including the step changes that are sometimes not immediately apparent in the degree of processing between sensor and display.

Figure 5: Progress of machinery control room design

6.2 CURRENT SITUATION

The current situation is that the control and monitoring of shipboard machinery involves a high level of complex, integrated digital systems that rely on embedded local modules, often invisible to the operator, and a plethora of sensors, actuators, control valves and other devices that enable operation that at one time would have required manual involvement. There is some evidence that electronic failures are regarded as 'normal' by engineers, who are principally trained as mechanical engineers, and these are shrugged off. In many cases the failures are not highly significant in terms of consequence. Failures are resolved by replacement of components that are, generally, low cost. Investigation of cause is not routine. Despite frequent similar failures, often involving the same component, automation is trusted. This may indicate that there is some mystique surrounding digital systems that results in a lack of challenge, possibly due to a feeling of helplessness. The inclusion of more education in electronics within all engineering courses would probably be beneficial.

The reasons for failures are many and various but it is likely that many are the result of electronic components, particularly sensors, being standard commercial off-the-shelf items that have not been subjected to a rigorous test programme to ensure that these are suitable for use in the maritime environment, especially in terms of resistance to vibration and tolerance of the temperature and humidity conditions that are normal in use. Potentially critical equipment, such as echo-sounders and speed logs have been found to show a mean time to failure, from new, of less than five years when used in temperatures above 25°C which is hardly abnormal for shipboard applications.

With the increasing volume of electrical and electronic components, planned maintenance systems have not evolved to reflect the changing maintenance requirements. Experience shows that the mean time to failure of, for example, a PLC is significantly shorter that the overhaul period for the mechanical parts. However, the necessary action to reduce the risk of unplanned outages is not likely to be included in any preventative maintenance schedule within the planned maintenance system. Additionally, with the rapid obsolescence of electronic components availability of replacement parts where the useful operating life can be problematic and will almost certainly come with a change of configuration involving a modification or upgrade.

The complexity of the digital control systems, including the associated sensing and monitoring, probably makes use of these error-inducing. Diagnosis of the information presented to the operator may be difficult to assimilate, with multiple alarms presenting simultaneously alongside nuisance situations, and a challenge in differentiating potentially hazardous conditions.

Without any doubt the sensory capability of the person is underutilized and the safe operation relies on the automation, which probably has not been designed with sufficient integrity and resilience to match the actual use. Systems are not designed, alongside the capabilities of people who will operate and maintain these, to deliver a total capability that matches the demand.

The incident history does not show the picture that has been found through anecdote. There have been a few major incidents that demonstrate a conflict between the capabilities of the automation and the people, generally attributed to a lack of sufficient knowledge and/or understanding of the, usually, new system. The generally-held belief of ship managers and seafarers is that there are far more 'incidents of interest' but reporting is incomplete.

One incident that probably merits further analysis is the contact between the large container vessel *Savannah Express* and the quay at the port of Southampton following a failure to reverse the main engine [26]. The comments by Duffey and Saull [8] in relation to the change on cockpit design would suggest that the industry could have expected incidents of this nature with the change to the electronic diesel engine, although few have resulted in major consequences. In this case the Chief Engineer had misunderstood the role of critical sensors, and following a number of failures and a lack of spares had continued operation without recognising the risks. The incident was preventable with hindsight but, following the argument of Bainbridge [24], the changes in the level of automation had increased the level of knowledge and understanding required in the operators. Training had failed to keep up with the demands of technology, as reflected in the recommendations of the investigators.

6.3 SENSORS

Following on from the case of *Savannah Express*, which was essentially the consequence of single sensor failure, and recalling the evidence cited in relation to Table 1, it is important to move from considering automation in general terms and turn to one of the weakest links – sensors. DNV GL has produced a useful paper [27] that addresses many of the issues faced as the industry increases the level of automation and adds new capabilities requiring more sensors, and in some cases the ability to detect different parameters requiring novel sensor technology.

Sensors have become complex systems in their own right, including signal processing elements, but at the same time these devices have been miniaturised. Some sensors are used in applications that are not critical but others, as was discovered in the incident involving *Savannah Express*, have the potential for causing the complete functional failure of a ship. Most sensors are micro-electronic devices and these exhibit the well-understood bath-tub pattern of through life failures. Some will escape the in-build tests and checks, only to fail at an early stage. The remainder will demonstrate a low, more or less constant, failure rate until at some point 'old age' brings an increased failure rate. The evidence available suggests that the useful life of many critical sensors in use is only a few years, and certainly less that the life of the items that rely on them.

There are approaches that can reduce the vulnerability of large systems to failure of single sensors. It is possible to build arrangements with, for example, three identical sensors and to feed the output signal through a voting system, so that the most probable value is used. It is possible to pursue the path of diversity and to use sensors of different types to give a, possibly, better chance of detecting a critical condition – using two types of fire detectors for example. The current position is that ships are in operation where failure of an individual sensor can cause the automatic shut down or slow down of the main propulsion. In many cases the crew will be able to replace the failed item and get the ship back on its way but if this happens, and it will give no warning, in a place where the safety of the ship is jeopardised a major incident could easily be the result.

Given the number of sensors fitted, and the variety, it is essential that the ship carries sufficient spares for those that are most critical, and which can be replaced by the crew. The suppliers of equipment should, during the design process, determine the criticality of all sensors. The criticality will inform the quality and integrity requirement for each device. The assessment will also provide a sound basis for the ship operator to determine which items should be carried as spares. The evidence suggests that the designers, or maybe those who are tasked with procuring the parts, do not fully understand the operating environment and the result is a high rate of sensor failures, and equipment stoppages. Cost will, no doubt, be a consideration but it appears that the balance is not correct between first cost and the in service burden of continual replacement. The author gets the impression that maybe some safety critical sensors should be routinely replaced at prescribed intervals, like fluorescent tubes, but the data to support such action is undoubtedly lacking.

When an exhaust valve position sensor, which must be operating in a fairly harsh environment, fails and causes a main engine slow down the scale of the challenge becomes apparent. The immediate need is to collate data, ideally with information on service life and details of the failure, to establish the nature of the challenge – even anecdotal evidence would be helpful – so that the issues can be explored. Training courses should be able to prepare seafarers for dealing with the associated problems, which will be encountered more frequently than the major breakdowns of the past.

6.4 FUTURE DIRECTION

The implication in this paper is that the only new technology worthy of comment is automation, however that is defined. The reality is that accommodating other new technologies in maritime operations can only be achieved through ever more sophisticated control and instrumentation systems. To achieve optimised performance these systems will require some level of automation, perhaps better described as autonomous, capability as it is no longer practicable for human operators to act as integrators, optimisers and managers of the complex integrated systems onboard. Although autonomy is often portrayed as a new frontier there is a long history of using autonomous devices on ships. After all, the governors of steam and diesel engines continually adjust to maintain the set conditions without any intervention, progressing from mechanical through hydraulic to electronic devices.

The addition of a number of different prime movers into a shipboard energy system – solar photovoltaics, Flettner rotors, various sails and kites, fuel cells – will only deliver the best energy-saving outcome if the different characteristics of each device can be combined with the others effectively, and that inevitably involves sophisticated digital power management systems. Integration of systems to gain 'better' performance will always come with greater complexity.

As technical complexity looks inevitably on the increase it is easy to forget that organisational complexity can add to the burden of the people who are tasked with operating the asset. As the scale of automation progresses, and that includes more genuinely autonomous elements, some of which include some form of machine learning, the challenges change. Ships that leave the yard as sisters, with an apparently identical design could become different as software 'learns'. Could that lead to problems for operators? The speed of advance means rapid obsolescence and unavailability of identical spares. Upgrading of software is not always properly managed and controlled. Configuration control needs to be improved.

Notwithstanding the challenges, the future will see a growth in software-intensive ships, controlled by a wider community of operators dispersed in several locations. Yet some of the hazards remain unchanged, and uncertain – weather, for example. The power of digital technologies can bring an exciting future to reality but there is a catch, referred to by Woods and known as Doyle's Catch [28]:

"Computer-based simulation and rapid prototyping tools are now broadly available and powerful enough that it is relatively easy to demonstrate almost anything, providing the conditions are sufficiently idealised. However, the real world is typically far from idealised, and thus a system must have enough robustness in order to close the gap between demonstration and the real thing".

It would be easy to conclude that the march of new technology, and the current fascination with unmanned ships and autonomy, is taking the industry along a pathway from adequate to good to better to best. The pathway becomes dangerous if the people are not kept in step with the changes.

7. BALANCING RISK AND BENEFIT

The previous sections are, inevitably, a mixture of gloom and hope. The safety record suggests that the long-term continual improvement has stalled in a time of rapid change. The industry is still struggling with challenging trading conditions that have prevailed for a long period of time.

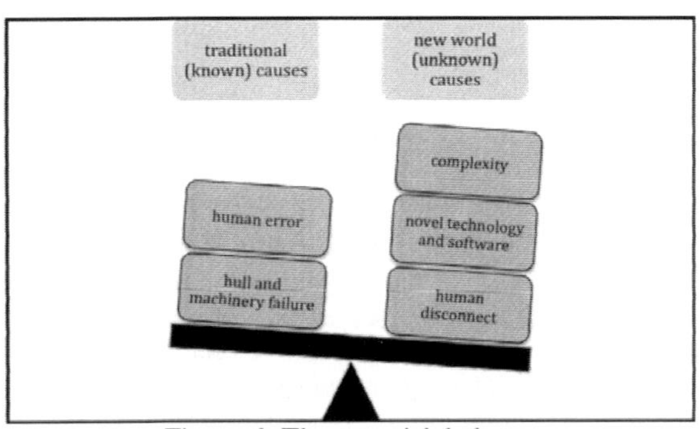

Figure 6: The new risk balance

Arguably, the regulatory regime is still focused on the traditional risks, as illustrated in Figure 6. Although there are still casualties resulting from unexplained, or inherent, hull and machinery failure the author subscribes to the view that most exhibit a considerable influence from the human element. Attempts to reduce human error are focused on the crew, and presumably for remotely-controlled ships this would transfer to the shore-based operators as the direct equivalents. The risks associated with the 'new world' in Figure 6 are the stuff of hushed conversations in dark corridors. There is evidence that the causes are real but there is nothing conclusive about the real world consequences beyond some anecdotal commentary and suspicion. Just consider the invasive nature of software that is at least safety-related – the early automation, or more correctly remote monitoring and control arrangements, was pneumatic, hydraulic or hard wired and the logic could be followed on a drawing and once completed it could be tested exhaustively. Once software is introduced the level of testing and examination has reduced. It has become impossible to test modern control system software exhaustively and completely, and the maritime industry does not have a good record of managing configuration and confirmation testing either. Even in industries that are considered to be even more reliant on software, such as aviation, there have been issues – for example the failure of generators on Boeing 787 Dreamliners after a number of running hours as data buffers became full [29]. When a modern electronic diesel engine might incorporate 2.5 million source lines of code and the bridge console, including ECDIS, over 3.5 million source lines of code the scale becomes apparent. To place these figures in perspective the avionics and flight support software for the Boeing 787 requires about 6.5 million lines of source code. By this comparison the maritime industry is not a simple one.

The risks associated with disconnecting the human through technology are less well understood, lots of anecdote and some commentary in investigation reports. The latter, and the number of cases seems to be increasing possibly due to the investigators developing a better understanding of the issues, suggest that whilst some of the lessons are best dealt with through training of watchkeepers there must also be a considerable need to include a more human-centred design philosophy into product development and system integration. It is also clear that some of the 'errors' made in taking information from modern displays are probably not unique to the equipment involved in a particular incident. The human factors community can make use of the limited number of cases of interest to inform design of all information displays to reduce the risk of misinterpretation and over-reliance.

The maritime industry needs to keep moving forward and that means embracing new technologies. The reduction of pollution of the planet through ballast water management and the control of exhaust emissions can be regarded as adding a burden in order to satisfy the expectations of society, in effect a charge imposed in order to stay in business. The changes are adding more functionality to the ship. The benefits mainly accrue to others, in terms of the control of the spread of invasive species and the improvement of health, but the risks fall to the shipowner, in terms of regulatory compliance and the more generic risks associated with owning and operating complex plant. New technologies are required and these should not come with unexpected detriment.

The most significant choice immediately facing the community is the selection of energy source, focused by the need to reduce the emission of carbon dioxide from fossil fuels. Whilst the benefits of moving away from consumption of fossil carbon will be at a societal level, shipowners also benefit as opportunities appear that could reduce operating costs and possibly even capital costs. The change could produce some radical solutions. The energy supply market is volatile and it is likely to become more dynamic as investment to reduce dependence on fossil fuels intensifies, and technology will probably be pulled through from non-marine sectors. There will be risks associated with marinisation, or failure to recognise the special operating conditions at sea.

The drive for change, and the corresponding need for new technical solutions requires a better understanding of the effectiveness of strategies for managing risk. When considering a stable requirement there are well established processes for assessing the risks associated with hazards and for managing these. The possible risk reduction strategies are shown on the left-hand triangle in Figure 7. When considering new technologies these are generally being introduced to manage the hazards (which can be very broadly defined, even at a societal level) and the questions shown on the right-hand triangle become the defining ones.

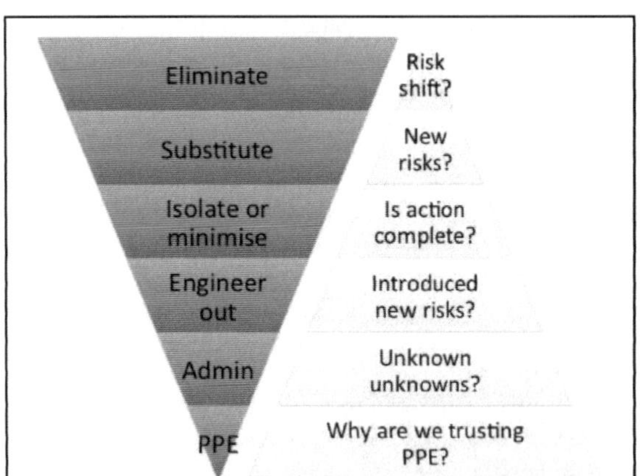

Figure 7: Changing technology, changing risk?

Woods [28] makes some very useful observations about the cross-industry fascination with autonomy – in some cases full autonomous operation as in road vehicles, in others remotely-controlled vehicles – and cautions that enthusiasm trumps thorough and comprehensive assessment of risk. He suggests that *"the effects of future technology change are unspecified, ungrounded and overconfident, whereas new risks are missed, ignored or downplayed"*. The focus is often on the technology and the impact of the human element, throughout the life-cycle is an afterthought. The maritime industry cannot afford to deploy new technology that increases risk.

As noted above the maritime industry has a preference for building on proven experience and, therefore, looking into the rear-view mirror. There is a history of dismissing suggestions that hazards exist where there is no evidence in the historical record. Even where there is evidence, it has been noted that hazards are dismissed because they should not have existed, typically where procedures have not been followed – the evidence is available that people actually behave like this and yet that hazard is not worthy of serious consideration. The IMO introduced the procedures for Formal Safety Assessment so that risks and benefits could be balanced, so that risk control options could be considered on a cost-benefit basis and to provide a structured approach to allow hazards, and associated risks, that have not been experienced to be evaluated. As technology changes this methodology would seem to offer some major benefits.

Maritime operations are full of risks and uncertainties. A concept was described by Brown [30], a lawyer, that suggested it was a reasonable expectation that in making decisions in relation to construction contracts the parties should take into account 'reasonably foreseeable uncertainty', that is if 'uncertainty' is 'reasonably foreseeable' there is no excuse to ignore that fact and to treat any occurrence as an 'Act of God'. That seems an interesting concept when thinking about the human element and unfamiliar technology or unfamiliar relationships with technical solutions.

8. CONCLUDING REMARKS

Shipping operates in a hostile physical environment and there are, and probably always will be, incidents that are the result of bad weather, and weather forecasting may improve further but it will never be perfect.

Once the human operator has been taken away from direct functional control there is a natural progression that moves through control stands, to control rooms with a direct view of the installation (or most of it), to control rooms nearby but without a direct view (and some owners opted for control rooms without windows onto machinery spaces to protect the crew from possible injury) to control rooms somewhere else entirely. The additional requirement included at each stage is a reliable communication highway to transmit data and control inputs. The contribution of the human element to safe operation shifts as this extraction from the traditional manual operation progresses. The early life-cycle phases of concept, specification and design take on much more significance since decisions and choices in those phases have a huge knock-on impact. These phases rely on the feedback from experience, whether encapsulated in rules and regulations or in less structured communications, to inform the designers. Where new technical solutions are being developed it is too easy to dismiss previous experience as 'no longer relevant' and existing rules and regulations as 'meaningless'. Shipping does not need to repeat mistakes of the past simply because the technology has changed. The industry needs to support the future developments by collecting data from operation and making it available.

New technology is not necessarily the best route to solving yesterday's problems and it has to be recognised that the human element contribution to the incident record has never been fully understood. New technology will always be required to solve problems in the best way possible. The adoption of new technical solutions should not be resisted, but these should be carefully assessed to ensure that the maximum learning is extracted from experience to manage the risks. Meanwhile, there are still gains to be made by focusing on the interaction between people and ships, wherever the people are situated, through good human-centred design, effective education and training and human-focused management practices.

9. ACKNOWLEDGEMENTS

The author gratefully acknowledges that work of the late John Crilley, a former colleague at Lloyd's Register, who began the systematic analysis of casualty data. The assistance of Barbara Jones and Anne Cowne of the Lloyd's Register Foundation Heritage and Education Centre in providing continued access to casualty records is greatly appreciated.

The author acknowledges the contribution of members of the Human Element Working Group of the Institute of Marine Engineering, Science and Technology to a wide-ranging discussion of issues relating to the interaction between technology and people.

10. REFERENCES

1. World Casualty Statistics, formerly Casualty Returns, published initially by Lloyd's Register, then Lloyd's Register-Fairplay and later IHS-Markit, 1980 – 2017.
2. PERROW, C., "Normal Accidents: Living with High-Risk Technologies", *Princeton University Press*, 1999.
3. DUFFEY, R.B. and SAULL, J.W., "Know the Risk: Learning from Errors and Accidents: Safety and Risk in Today's Technology", *Butterworth-Heinemann*, 2003.
4. BAILEY, N., ELLIS, N. and SAMPSON, H., "Training and Technology Onboard Ship: Hoe seafarers learned to use the shipboard Automatic Identification Systems (AIS), *Lloyd's Register Educational Trust Research Unit, Seafarers International Research Centre, Cardiff University*, 2008.
5. TURBERVILLE, C., quoted in "Shipping and Safety Review 2017, *Allianz Global Corporate and Specialty SE*, 2017.
6. "Shipping and Safety Review 2018, *Allianz Global Corporate and Specialty SE*, 2018.
7. LOUGH, A. and POMEROY, R.V., "The Millennium Bug – Technical issues in relation to marine systems", *in Guidelines for Year 2000 projects: A guide to solving the problems associated with ships and the millennium bug*, Lloyd's Register, 1998.
8. POMEROY, R.V. and EARTHY, J.V., "Merchant Shipping's Reliance on Learning from Incidents – A habit that needs to change for a challenging future", *Safety Science*, 2017.
9. "Safety and Shipping Review", *Allianz Global Corporate and Specialty SE*, 2013, 2014, 2015, 2016, 2017 and 2018
10. "Human Element Vision, Principles and Goals for the Organization", *Resolution 947 adopted at the 23rd Session of the Assembly, IMO*, 2004.
11. "Grounding of the Panamanian Passenger Ship *Royal Majesty* on Rose and Crown Shoal near Nantucket, Massachusetts June 10, 1995", *National Transportation Safety Board, NTSB/MAR-97/01*, 1997.
12. "Report on the investigation of the collision of *Rickmers Dubai* with the crane barge *Walcon Wizard* being towed by the tug *Kingston* in the south-west lane of the Dover Strait Traffic Separation Scheme on 11 January 2014, *Marine Accident Investigation Board, Report No 29/2014*, 2014.
13. "Passenger Vessel *L'Austral*, grounding, Milford Sound, Fiordland 9 February 2017, *Transport Accident Investigation Commission, MO 2017-202*, 2018.
14. "Statistical Data on Failure of Steering Systems, Propulsion Plant and/or Electrical Plant and Reliability and duplication of Ship Components: A statistical approach to the reliability of equipment fitted on board ships: Note by the Secretariat", *Paper DE XXIV/8/1 presented at the 24th Session of the Sub-Committee on Ship Design and Equipment, IMO*, 1981.
15. "Statistical research on failures of ship propulsion systems", *Paper DE/321 presented as a background document at the 26th Session of the Sub-Committee on Ship Design and Equipment, IMO*, 1982.
16. POMEROY, R.V., "The Generation and Use of Reliability Data for Marine Engineering Systems", *Fourth International Symposium on Marine Engineering Systems*, ICMES, 1987.

17. "Catastrophic engine failure, resulting in a fire and serious injuries to the engineer on board *Wight Sky*, off Yarmouth 12 September 2017", *Marine Accident Investigation Board, Report No 14/2018*, 2018.
18. Van der Vleuten, E., Oldenziel, R. and Davids, M., "Engineering the Future, Understanding the Past: A social history of technology", *Amsterdam University Press*, 2017.
19. DUFFEY, R.B., private communication with author.
20. "Guidelines on alternative design and arrangements for SOLAS Chapters II-1 and III", *MSC.1/Circ.1212, IMO,* 2006.
21. "Automation in Ships: Note by the Secretariat", *Paper MSC VIII/11 presented at the 8th Session of the Maritime Safety Committee, IMCO*, 1964.
22. MUNTON, R., McNAUGHT, J. and MACKENZIE, J.N., "Progress in Automation", *Transactions of the Institute of Marine Engineers, Vol 75*, 1963
23. MUNTON, R. and McNAUGHT, J., "Automation of High-powered Diesel Machinery", *Transactions of the Institute of Marine Engineers, Vol 78*, 1966
24. BAINBRIDGE, L., "Ironies of Automation", *Automatica, Vol 19*, 1983.
25 "Report on the investigation of the impact with the quay by the passenger ro-ro ferry *P&OSL Aquitaine* at Calais on 27 April 2000", *Marine Accident Investigation Board, Report 27/2001*, 2001.
26. "Report on the investigation of the engine failure of Savannah Express and her subsequent contact with a linkspan at Southampton Docks 19 July 2005", *Marine Accident Investigation Board, Report 8/2006,* 2006.
27. "Understanding Sensor Systems Reliability", *DNV GL Technology and Research Position Paper 2-2016,* 2016.
28. WOODS, D.D., "The Risks of Autonomy: Doyle's Catch", *Journal of Cognitive Engineering and Decision Making, Vol 10.* 2016
29. "Generator bug prompts 787 directive", *Flight International 5-11 May 2015*, 2015.
30. BROWN, N.A., "Reasonably Foreseeable Uncertainty", *Pinsent Masons*, 2011.

11. AUTHORS BIOGRAPHY

Vaughan Pomeroy worked for ten years in the aircraft industry and with an international practice of consulting engineers before joining Lloyd's Register in 1980. He worked in research and engineering design appraisal before being appointed Deputy Chief Engineer Surveyor in 1992. He later managed research and development for the marine business, playing a key role in the development of naval ship classification and in the initiation of the Common Structural Rules for oil tankers and bulk carriers. He was appointed Technical Director, before retiring in 2010. He was appointed as Adjunct Professor by the University of Southampton in 2004 and continues in this appointment, doing some teaching and individual research.

SOME ERGONOMIC ISSUES OF DP VESSEL CONTROLS

M Hadfield, Health and Safety Executive, UK

SUMMARY

A number of loss of position incidents have occurred due to accidental, unintentional, transfer of control from DP to manual. The incidents can be attributed largely to poor ergonomic design of the DP control systems.

A particular incident with a drilling rig is described and the weaknesses in the ergonomics discussed.

Similar situations have been found on other vessels. The measures being taken, or considered, by the industry and regulators are described.

1. INTRODUCTION

The Health and Safety Executive, is a UK government department. Amongst other duties, it is the regulator for health and safety in the UK offshore oil industry. It became aware of a loss of position incident on a drilling rig in 2016 and participated remotely in the Drilling Company's investigation. HSE formed the view that it was largely due to poor ergonomics.

A Safety Alert to advise the industry was issued. HSE at the same time, became aware of two very similar incidents worldwide. Since then more have come to light.

The main ergonomic weaknesses were a) inadequate protection against accidental transfer of mode of control from DP to manual and b) no clear indication that mode of control had changed from DP to manual.

The situation on other vessels is discussed together with actions taken by regulators and others.

2. INCIDENT

2.1 SUMMARY

A semi submersible DP drilling rig lost control of position for several minutes while it was connected to a well by its drilling riser pipe. During this time, it was obliged to shear the drill pipe (which passes through the riser and down into the well) and disengage the lower marine riser package, LMRP. The initial loss of control was due to accidental disengagement of the DP system. The crew immediately noticed loss of position but did not appreciate that DP was disengaged. They initially believed there was a technical fault with the DP and it took 6 minutes before they realised the DP was disengaged.

Both the loss of position control and the inadequate initial crew response were largely due to poor ergonomic design of the control system. Firstly, the button for transfer from DP to manual control was not protected against accidental operation. Secondly, there was no clear indication at the DP desk that DP was no longer engaged, and that the vessel was then under manual control.

2.2 CAUSE

The Company's investigation found that the initiation of the incident was accidental, unknowing, operation of the button used to transfer control from DP to manual. (A dual push was required but this gave inadequate protection in this instance.)

2.3 RESPONSE

The DP Operator on the DP desk thought that DP was still engaged and continued to operate at the DP control desk and tried to find a technical fault. There was no clear indication at the DP desk that control had passed from DP to manual lever mode. (The utility panel for selection of DP or manual mode was two desks away from the operational DP desk, and there was no clear indication on his screen.)

As is normal practice, there were two DP operators on duty. Typically, one monitors and adjusts (if necessary) the DP set up for a period of 2 hours while the other is engaged in other bridge work. They then swop over.

If a drilling vessel loses dynamic positioning it will move off its designated location. The riser can accommodate movement within certain limits, but if these are exceeded the riser tensioner system, or other parts of the riser package, will reach the end of its travel and serious damage can result either on the drilling rig or the subsea well. To avoid this, it is normal to make an emergency disconnect. This requires the driller to shear the drill pipe with the shear rams in the seabed blow out preventer, BOP. (He must first adjust the height of the drill string to ensure that there is not an unshearable drill pipe joint in way of the shear ram.) Valves on the seabed BOP and the bottom of the riser, LMRP, are then closed and the LMRP is disconnected from the BOP on the top of the well. The riser is then hanging freely below the drilling rig as it drifts off location. Depending on all the circumstances, if this is not done before the riser limits are reached, it may prove impossible to perform some or all of these actions. In the worst scenario, for some situations, oil and gas could reach the drilling rig or sea surface and threaten the installation.

An out of position alarm is raised once a 'warning circle' is reached. Everyone then knows that unless position control is regained, an emergency disconnect must be affected and that there may be very little time to do so, depending on how quickly the rig is moving off location. An emergency disconnect in itself has high cost consequences. Recovery can take many days. The rig must be repositioned and reconnected to the well – assuming the well owner is prepared to allow the rig to continue. The drill and sheared drill pipe left in the well must be recovered before a new drill string is made up to continue the work.

In this highly pressured, stressful period, the bridge team's effort is divided between trying to regain position control and preparing for emergency disconnect by keeping the driller (situated near the drill floor) informed. An added complication in this incident was that a supply vessel was connected and bunkering fuel to the rig at the time. The bridge crew had to liaise with this vessel and arrange its disconnection.

The offshore installation manager, OIM, of course came to the bridge at this time.

In the event, after 1 minute an alert was issued to the drill floor, at 3 minutes, the 'red' limit of 8 metres off location was reached and the emergency disconnect sequence commenced. At 5 minutes successful emergency disconnect was achieved as the LMRP was unlatched from the BOP.

At 6 minutes it was realised that vessel mode of control was in manual, and the bridge team was then able to control the position of the vessel.

2.4 ERGONOMIC FACTORS

The control desks and layout are shown in Figures 1,2 and 3. During the incident, the middle DP desk was the operational desk at which the DP operator was working.

The buttons to select mode of control are in a small utility panel on the riser management desk which was two desks to the operator's left. Fig 3 shows the DP selector button illuminated to show DP is selected. The adjacent button would select manual lever control, and that button would then be illuminated.

Two main ergonomic factors contributed to this incident:

2.4(a) Inadequate protection against accidental change of DP status – mode of control

The only protection against accidental activation was a double push requirement on the button. This required two pushes within 2 seconds.

Better protection measures are given in American Bureau of Shipping Guidance, Ref 1. Section 2 Controls 2.7 Prevention of Accidental Activation states:

> Controls should be designed and located so they are not susceptible to accidental activation. Methods to reduce the likelihood of accidental activation include:
>
> i) Locating and orienting the control so that bumping is unlikely to cause activation
> ii) Providing sufficient control resistance to prevent unintentional movements
> iii) Requiring complex motions for control activation, such as an interlock or rotary motion
> iv) Restricting access to controls by isolating them or by providing a cover guard or physical barrier

2.4(b) Inadequate Indication of Status or Mode of Control

There was no clear indication at the DP control desk that DP was disengaged and the vessel was now under manual control. The desk in use was the middle desk – OS#2 in Fig1.

The main message on the DP screen was 'THRUSTER NOT READY'. In the circumstances this is not helpful. It could indicate a variety of situations. It certainly does not convey the fact that the mode of control had switched from DP to manual.

(DP screens can be very busy containing much information. There may have been a small display somewhere on the screen from which change of mode of control could be inferred.)

The most obvious indication would have been the light on the lever manual button on the utility panel which would have become lit rather than the adjacent DP button. However, that is not actually very obvious being some distance from the then operational DP desk and (it is believed) both lights were the same colour.

The equipment did not comply with Industry guidance. For instance, the ABS Guidance, Ref 1, Section 3 2.4 'Status Information' states:

> - Visual displays should provide a positive indication of the state of the equipment such as: ready, running, not running, etc.

Similarly, the IMO Circular, ref 2, Section 3.4.1.3 states:

- The active (control) mode should be clearly displayed.

2.4(c) Other Ergonomic Aspects

The investigation found that it was likely that the manual button had been accidentally activated when someone was working at the Riser Management desk. It might have been leant on or notes might have been taken above the button. Possibly papers or documents needed for this work had been placed on top of the button. It is recognised bad practice to place anything onto DP desks in this manner. However, the equipment designers do not seem to have provided a suitable place for such papers.

The company also found that the audible alarms did not comply with good practice on alarm management. This aspect is not considered further in this paper.

2.5 CORRECTIVE ACTIONS TAKEN

Soon after the incident, a cover guard was fitted to the utility panel as shown in Fig 3.

Other actions taken by the company included a ban on objects, including paper, being placed on DP consoles, increased DPO participation in Emergency Drills, and such drills to include unexpected change from DP to Manual mode.

3. OTHER INCIDENTS

In 2016, NOPSEMA, the Australian regulator, and the US Coast Guard issued safety alerts about very similar incidents relating to a Dive Support Vessel and a Drill Ship respectively. The alerts are listed as Refs 4 and 5.

IMCA, the International Association of Marine Contractors, has issued a 2018 Bulletin, Ref 6, saying it was aware of 9 similar incidents for the three years, 2015 thru 2017, of which 6 were with double push button and 3 were with single push button.

4. HSE EXPERIENCE

Since the 2016 incident, this topic has been inspected when HSE Maritime Integrity inspectors have visited DP installations.

It has been found that most of the vessels originally had no protective covers over the control mode selector switches. (Switches may be button or other type.) About half those visited had at some point fitted protective covers. Where no cover was present, the owner was required to fit one. To date this has been done willingly.

Such covers are normally held onto the control desks by adhesive, because screwing or bolting them in place could jeopardise the warranty on the equipment.

It has been found that DP systems from the incident's manufacturer have the same weaknesses in information display. There is no clear indication of the mode of control (manual or DP) at the DP screens, and the same alarm message 'THRUSTERS NOT READY' would appear. In comparison to fitting a protective cover, improvements to this aspect are less easily achieved because they require software changes and the active involvement of the equipment supplier.

On initial contact, the DP manufacturer felt there was nothing wrong with their systems and equipment.

It has also been found on a number of vessels that there was no suitable place for papers, checklists or notes etc, to be positioned. In one case a shelf had been fitted by the crew on the side of the bridge chair for this purpose.

It seems likely that either ergonomic expertise was not involved in the design of the control desks by the manufacturers or the original customers or it was somewhat ineffective.

5. WIDER INDUSTRY RESPONSE

At the 2017 IRF, International Regulators Forum, AGM in Denmark, NOPSEMA, with HSE backing, presented the latest information on the risk posed by design-induced human error for DP systems. The offshore industry regulators agreed to pursue these issues in their jurisdictions.

Additionally, NOPSEMA has delivered the presentation at industry conferences and included a number of articles in their 'Regulator' magazine. They also wrote to global DP manufacturers seeking information on the steps they have taken regarding the tolerance of their systems to human error. From the four that responded there was a general recognition that double-press buttons are susceptible to human error.

One DP manufacturer issued a DP Information Letter in March 2018. This states that for decades the double push button has been an accepted means to act as a safety barrier against accidental deselection of DP mode. Several vessel owners have requested the company to provide an additional barrier of either a protective cover or a software function providing screen based pop-up dialogues. (These would require confirmation of deselection of a control mode.) The manufacturer prefers the software option and considers that the decision to install an additional barrier is the owner's responsibility.

6. CONCLUSION

The incident described, plus several others, would probably not have happened if there had been better protection against accidental deselection of the DP mode of control.

There is increasing recognition that a double push button does not by itself provide adequate protection. In the offshore oil industry, many installations and vessels have now fitted protective covers over the mode selection switches. This is one of the recognised protective measures in American Bureau of Shipping guidance.

Effective response to the loss of position was delayed, because, contrary to industry guidance, there was no clear indication on the DP screen of the current mode of control – which would have indicated that control had transferred from DP to manual. The author is not aware of improvements directly meeting this requirement. However, software modifications are becoming available which require transfer of mode of control to be confirmed before implementation.

Other ergonomic weaknesses present were inadequate locations for papers and documents, and unhelpful alarm arrangements.

It would appear that, at the design stage, Human Factors ergonomic expertise was either not employed by the manufacturer or the original customers or it was somewhat ineffective.

7. REFERENCES

1. 'Guidance Notes on The Application of Ergonomics to Marine Systems' American Bureau of Shipping 2013

2. 'Guidelines for Vessels with Dynamic Positioning Systems' International Maritime Organisation MSC Circular 645

3. 'Some Ergonomic Issues of DP Vessel Controls', Health and Safety Executive (HSE) Safety Alert Bulletin ED3 2016 http://www.hse.gov.uk/safetybulletins/dp-vessel-controls.htm

4. 'Vessel loss of position whilst diving in close proximity to a hydrocarbon facility' NOSPEMA (National Offshore Safety and Environmental Management Authority) Safety Alert 62 June 2016 https://www.nopsema.gov.au/assets/Safety-alerts/A484748.pdf

5. 'Lessons Learned: Dynamic Positioning Ergonomics' USCG National Centre of Expertise 'Drill Down' Issue 7-LL October 2017 https://www.dco.uscg.mil/OCSNCOE/The-Drill-Down/ in the listing item 7.

6. 'IMCA DP Station Keeping Event Bulletin 01/18' International Marine Contractors Association February 2018

8. AUTHOR'S BIOGRAPHY

M Hadfield has been a Specialist Inspector in the Maritime Integrity team of the Energy Division of the UK Health and Safety Executive since 2009. He deals with the marine integrity aspects of floating offshore installations through inspections, incident investigations, and safety case assessment.

9. FIGURES

Figure 1: Bridge View showing manual lever control desk, RMS riser management desk, and three DP control desks.

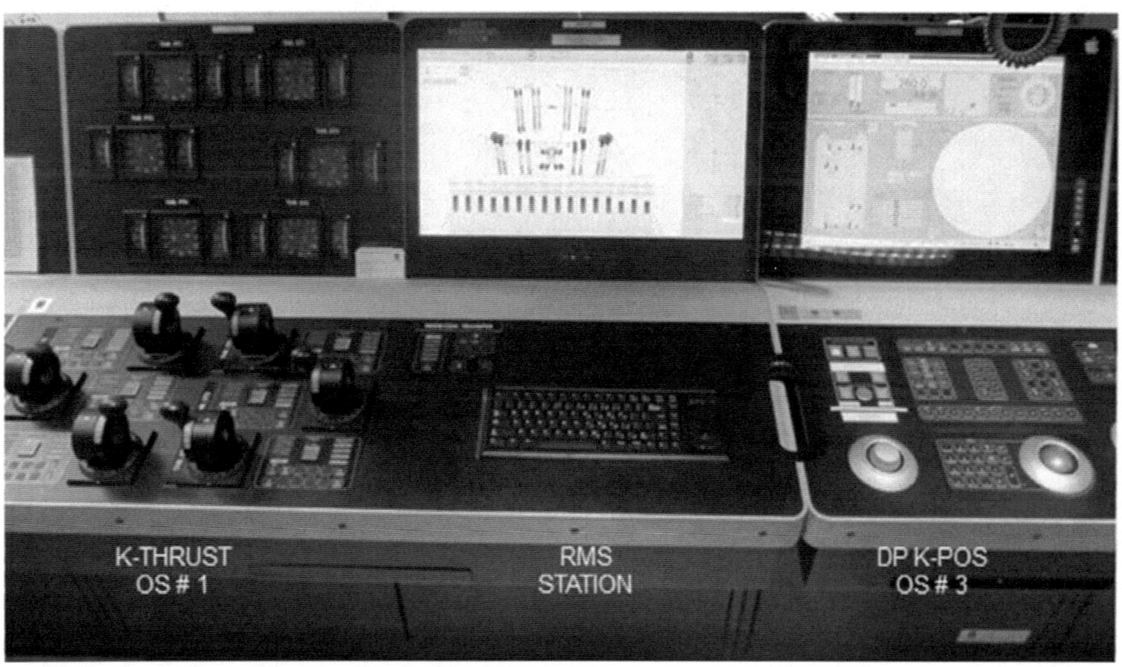

Figure 2: Close up of manual lever desk, riser management desk, and one DP desk. The utility panel is top left on the riser management desk

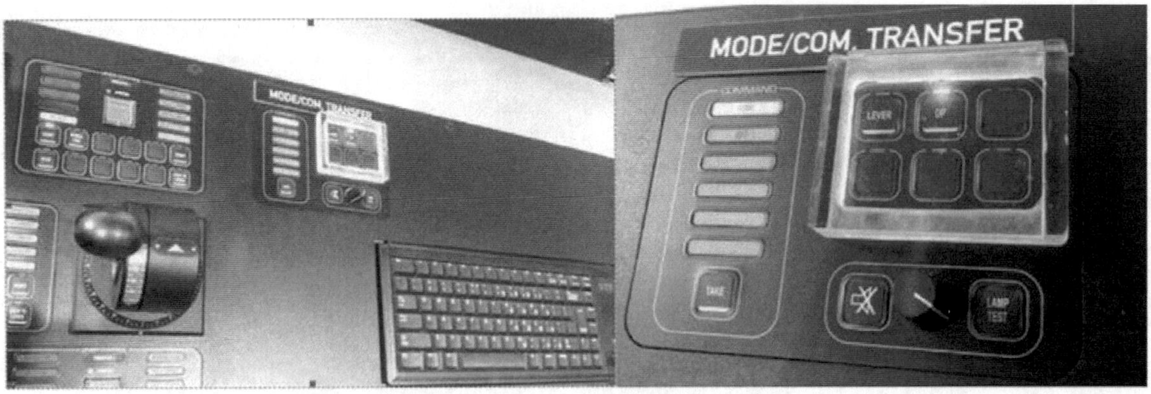

Figure 3: Close ups of (left) a manual lever control and the riser management desk, and (right) the utility panel with a protective Perspex cover fitted after the incident.

PRESENTATION

FROM STEM TO STERN; SAFETY STARTS WITH DESIGN

H.A. Oltedal, Department of Maritime Studies, Western Norway University of Applied Sciences
R. Wild, Harwich Haven Authority, Western Norway University of Applied Sciences
M Lützhöft, Department of Maritime Studies, Western Norway University of Applied Sciences

SUMMARY

A safety management system should create and support an organisational process based on individual learning from operations which discovers system flaws and weaknesses to prevent future failures. An efficient safety management system should incorporate knowledge about how misconceptions and poor design have contributed to failures. The maritime community should have mature safety management systems by now; but shipping companies repeatedly face the same risks and hazards. By using the example of *Hoegh Osaka* incident, we claim that a fundamental weakness of current safety management is that managers do not address the core nature of these risks. Human error is still far too often used as a causal factor and the safety measures are introducing new procedures and checklists, instead of improving the existing ones. We also argue that one critical factor is the lack of seafarers' participation in the management of the Safety Management System. By placing the seafarers in the centre when designing a safety management system we strengthen their authority, and support the human contribution to safe operations in a dynamic environment.

1. INTRODUCTION

The International Safety Management (ISM) Code, which was adopted by the International Maritime Organization (IMO) in 1993, requires shipping companies to implement and maintain a safety management system (SMS). One regulatory requirement of the ISM Code is reports and analysis of non-conformities, accidents and hazardous occurrences, which are regarded as valuable tools to prevent future accidents and failures under the assumption that different types of adverse events may share the same underlying causes.

Managing safety by learning from previous experience, failures and using feedback control is a traditional systemic approach towards safety management [1, 2], but has clear limitations in social non-deterministic systems as for example shipping. Adverse events appear in many forms with their causal paths influences by situational, social and technical characteristics which originate in the surroundings, the vessel itself, crew, working environment and more, and make it impossible to predict the future.

Stacey's theory of complex responsive processes of human relationship offers an alternative view towards organizational learning [3], under which the acceptance of uncertainty and unpredictability is fundamental as no one can foresee or control the interplay of intentions between social actors. This approach emphasizes the patterns of relationships and communication between people, and on understanding these processes in order to enable good decisions through good interrelationships. This is in contrast to the traditional view of safety management that emphasizes control.

When developing efficient control measures to increase safety it is essential to understand the locus of control. If an organisation believes they can control future actions with procedures, and assume that social actors always act as intended, they may over-emphasize the use of control measures as procedures and check-list. They must then design these procedures and check lists for every conceivable situation [4] in order to avoid, or correct, deviations from a wanted state in the belief that uncertainty can be removed.

This approach may result in an inefficient safety management system that does not have legitimacy with its users, and thus may degrade safety instead of increasing it.

On the other hand, the theory of complex responsive processes strives to understand and explain what actors are doing (work as done), rather than what social actors ought to be doing [5] – including the interplay between social actors, communication, relations of power and decision-making. This is an alternative way to design an organisational safety management system that support and empowers social actors to take the best possible decision in any situation,

In this article, we will critically discuss the traditional view towards safety management adopted by the IMO and the ISM code by using the example of the *Hoegh Osaka* incident. We suggest an alternative view that emphasizes giving the social actors – the seafarers – working conditions and procedures which facilitates making the right decisions.

2. SAFETY VS. COMMON SENSE - THE DICHOTOMY MUST CHANGE

2.1 NARRATIVE

On 3 January 2015, the pure car and truck carrier *Hoegh Osaka* was rounding West Bramble buoy in The Solent when it developed a significant starboard list causing some cargo shift and consequent flooding. With the list in excess of 40°, the ship lost steerage and propulsion, and subsequently drifted onto Bramble Bank, grounding. Stability modelling and analysis following the accident show that the *Hoegh Osaka* heeled heavily to starboard while turning as a result of having departed port with inadequate stability [6]. A key finding of the MAIB investigation is that no departure stability calculation had been carried out on completion of cargo operations and before the *Hoegh Osaka* sailed. The report also indicates that the fundamental requirements for establishing before departure that a ship has a suitable margin of stability for the intended voyage had been eroded on board the *Hoegh Osaka* such that unsafe practices had become the norm. Unsafe practices along with other variations of human error is said to be the principal cause in all – or most - incidents and accidents at sea [1]. In a feedback safety management system, safety measures address what investigations identifies as causal factors, and when this is identified as some form of human error, safety measures often come in the form of instructions, check-lists and procedures. This is also the case with the *Hoegh Osaka* incident, where MAIB made recommendations to further enhance the respective instructions and procedures.

2.2 TRADITIONAL SAFETY MANAGEMENT

The procedures on the *Hoegh Osaka* were governed by a series of checklists addressing prior load/during load, prior discharge/during discharge, prior (arrival) discharge port, loaded passage and during loading/discharge. Such checklists are an integral part of safety management today and are typical of all ships, operating under the International Safety Management (ISM) code. Some of the checklists had *Additional Notes* appended to them. Whilst it is true to say that there were more than 200 checks to be signed-off, as quoted in the investigation report, the lists cover very different activities and it would be inappropriate to complete then all simultaneously. Nevertheless, the number of checks to be undertaken and signed off for a single activity is considerable. For example, the Prior Loading/ During Loading Checklist No 1 have 48 items plus 12 Additional Notes as described above.

Following an analysis of the checklists (the annexes to the main report) a number of further points were identified by the authors. These are symptomatic of the over-dependence and proliferation of checklists that do not and will not prevent incidents. These points were not identified in the MAIB report:

1. The sign-off is by the Duty Officer and Master. We submit that the second signature of the Master is largely meaningless because a) there is no second check column and b) it is unrealistic that the Master undertakes a physical check due his other responsibilities, and therefore all the Master is doing is witnessing the Duty Officers' checklist.
2. The status of the Additional Notes within the checklists is unclear. It is not known why the notes are appended separately, if the checks can be delegated and there is some duplication. For example, Item 28 relates to the placement of protective Foam Pads, Additional Note 3 concerns foreign objects such as empty bottles. Item 16 relates to hold lighting and Additional Note 8 also concerns hold lighting.
3. On checklist 1 (prior/during loading), Item 10 (vessel alongside) and Item 12 (offshore winds) clearly relate to dynamic factors which can change at any time throughout the loading (or discharging) operation. A single sign-off is not the same as a record that these factors are being monitored on a regular basis.
4. Item 32 (provision of space) is actually a subsidiary checklist of 7 items (A to G) within checklist 1. The areas to be checked for access are distributed in a wide area of the ship: pilot walk, engine room and emergency exits (where?). There is no detail as to where some of these items are and there is a single sign-off box for all of them. We submit that without additional detail and sign-off, the natural tendency is to sign the checklist, covering all these 7 additional items without a proper physical check.

As identified in the investigation report, the checklists, for all their apparent comprehensiveness, were incomplete. For example, checklist 5 (during loading) required a check of ship stability, whereas Checklist 3 (prior departure port) did not require such a check. We submit that far from assisting staff with conducting a safe operation, such checklists encourage short cuts. A list of 48 items, with sub lists and additional notes, illogical structure and poor sequencing, hinders safety management because it is restrictive, static and irresponsive to dynamic change. The MAIB report on the *Hoegh Osaka* recommended that 'checklists should be revised so that safety critical items are not lost'. Furthermore, any checklists should reflect reality, and this includes items that arise from incidents [7, 8].

Dynamic change, that can cause a departure from a plan of intended work, often checklist-led, is a reality during many operations at sea. Reacting to change, even if all that is required is acknowledgement that there has been a change, can be critical. This is because, as identified by Loukopolous [9] a static checklist can be inadequate as soon as there is an interruption or deviation from the intended list of actions. There is a possibility that when, or perhaps if, the actor returns to the checklist, completion of it does not happen or steps are omitted.

2.3 FEEDBACK MANAGEMENT SYSTEMS

The traditional route to a safer operation is to record and analyse incidents that went wrong and then erect barriers to prevent the errors reoccurring again [1, 2]. Such barriers can be new procedures, new equipment or new training, a combination of all these and more. This method of managing safety is reactive and for most of the time it is successful. Indeed, after a significant incident such as the capsize of T*he Herald of Free Enterprise* in 1987, which was the precursor to the ISM Code, there is a period of complacency because the anxiety of a repeat incident would appear to have been addressed. However, the provenance and success of the ISM Code is attracting increasing criticism in studies focussing on incident causes [10, 11].

What is overlooked in a prescribed and post-hoc management structure is the number of operations or events that are concluded without ceremony, because they were successful. In effect, the mundane becomes the norm and no one takes notice. Whilst this can be a recipe for complacency [12], the execution of the successful tasks is a potential resource often ignored.

3. SAFETY MANAGEMENT REDESIGNED FROM THE SEABED UP

3.1 THE HUMAN CONTRIBUTION

A more resilient method of managing safety is to learn from the success stories and enhance individuals' ability to learn, monitor and adapt [13, 14, 15]. Organisations may have the wherewithal within themselves to become more resilient and a review of that is appropriate.

Organisations are/can be characterised by intangible artefacts such as culture and also community of practice. This is not new, and an illustration is the passing of skills and customs both horizontally (between peers) and vertically (between generations). The phrase 'community of practice' (CoP) encapsulates this notion and was coined in 1991 by founding theorists Lave and Wenger [16]. Within the ideas/themes/skills cascaded as described, there will be good and bad practice. A resilient operation should capture and preserve the good practice.

Any change to safety management must be from the bottom up. Clearly, there must be a drive to implement change today. But ultimately, a new approach must be instilled not only in those charged with designing systems and procedures, but also in those just embarking on a seafaring career. If the youngest seafarers are imbued with a different approach and good practice from the outset, a new community of suitably trained practitioners will grow and expand. This will take time to achieve, and no one should expect instant results and superior quarterly or annual safety statistics, that can be reported immediately.

Accepting that human error is said to be the principal cause in all – or at least most - incidents and accidents at sea is an oft quoted statement that must be challenged as it is too simplistic, so that a new design can emerge. This is because no maritime operation is carried out without the involvement of a human being, it will always be a human being involved, and when reports are being analyzed in the search for underlying cause(s), far too often the investigation stops when the direct cause is revealed with the label human error – which inhibit the underlying cause to be revealed. The human element and human error are different concepts which are often misinterpreted and used interchangeably as if they were the same.

Human error is a deviation from intention, expectation or desirability. Training to enable operators (in this case, seafarers) to manage a constantly changing and dynamic environment is critical. Checklists serve a purpose, but only up to a point. Operators must be trained to understand that deviations from 'work intended' can occur and they must understand how to cope. The safety system must be designed to promote and address these two points. Scenario testing, a common tool used to predict the future, can be used to develop an understanding of dynamic change that alters work intended to work undertaken. Once there is a tacit acceptance that 'work undertaken' will not always mirror 'work intended' – then, and only then, can safety management move forwards.

4. CONCLUSION

Dreyfus and Dreyfus [17] explain that when the human actor is governed by checklists and written rules, she is inhibited from developing the intuitive competence that is required to manage unexpected situations. Checklists and written procedures are more suitable for novices, however, only when they are designed in a way so they are perceived as logical and practical. Therefore, a safety management system should facilitate and foster human actors having the tools and skills to adapt.

Crew who have been a long time on a ship will amass silent knowledge about the system, the ship's condition, what works and what does not work, almost by intuition.

Crew stability has been shown to support safe practice [18]. Changing the whole crew may be a bad idea. Previous service on a similar ship can be useful but is not always a valid argument, as the system includes much more than the ship. Furthermore, even sister ships have many differences, due to age, updates, retrofits and maintenance variations.

If, as in this case, many changes occur regarding for instance sailing destination and cargo, the crew should be supported in the management of such changes in a timely manner, considering the consequences such changes may have on the system. One example is that crew members could be trained in collecting and accessing information regarding previous good practice. This can be enabled via a range of methods, during hand-over meetings and periods. This will facilitate experience transfer in order to find the best possible solutions to manage the consequences of changes to the system.

To address (achieve) this, an alternative to written rules are to focus on organisational measures such as crew stability to reinforce trust and knowledge, and training that promote group discussion and imagination in order to handle an unknown future state and possible consequences of operational changes. Thus, instead of trying to control the future and human actions by the means of written rules – an approach that may turn out to be inefficient – the designers of a safety management system should reinforce the human role, in order to make them capable to manage unexpected changes in the operational system.

5. REFERENCES

1. REASON J. Managing the risks of organisational accidents. New York: Ashgate Publications Ltd.; 1997.

2. HOLLNAGEL E. Barriers and Accident Prevention. Aldershot, UK: Ashgate; 2004.

3. STACEY R. Complex Responsive Processes in Organizations. New York: Routledge. 2001.

4. DEKKER SWA. Follow the procedure or survive. Hum. Factors Aerosp. Safety; 1(4): 381-385. 2001.

5. BRAITHWAITE, J, WEARS, RL, HOLLNAGEL, E. Preface: Delivering Resilient Health Care. In: Resilient Health Care, Volume 3: Reconciling Work-as-Imagined and Work-as-Done. Jeffrey Braithwaite, Robert L. Wears, Erik Hollnagel (eds). Boca Raton, CRC Press. 2017.

6. MAIB. Report on the investigation into the listing, flooding and grounding of Hoegh Osaka Bramble Bank, The Solent, UK on 3 January 2015. Southampton: Marine Accident Investigation Branch. 2016.

7. WILD R, CONSTABLE K. A Document Of Debatable Value – A Case Study Into The Use Of Master-Pilot Exchange Documentation In Selected UK Ports. Journal of Navigation 66 (3), 465-471. 2013.

8. WIENER, E. Intervention Strategies for the Management of Human Error: Technical Report, NASA; United States. 1993.

9. LOUKOPOULOS LD, DISMUKES RK, BARSHI I. The Challenge of Routine Flight Operations. Proceedings of EAM 2003. The 22nd European Conference on Human Decision Making and Manual Control June 2-4, Linköping University, Linköping, Sweden. 2003.

10. BATALDEN BM, SYDNES AK. Maritime Safety and the ISM Code: a study of investigated causalties and incidents. WMU JoMA Vol 13. 2014.

11. BHATTACHARYA S. The effectiveness of the ISM Code; a qualitative enquiry. Marine Policy 36 (52), 528-535. 2012.

12. ÅRSTAD, I. and T. AVEN. Managing major accident risk: Concerns about complacency and complexity in practice. Safety Science 91, 114-121. 2017.

13 HOLLNAGEL, E. How Resilient Is Your Organisation? An Introduction to the Resilience Analysis Grid (RAG). Available from https://www.researchgate.net/publication/281803346_How_Resilient_Is_Your_Organisation_An_Introduction_to_the_Resilience_Analysis_Grid_RAG. 2015.

14. PATTERSON, M. and E.S. DEUTSCH. Safety-I, Safety-II and Resilience Engineering. Current Problems in Pediatric and Adolescent Health Care 45 (12), 382-389. 2015.

15. EUROPEAN COMMISSION. From Safety I to Safety II: A White Paper. Bruxelles: EC. 2013.

16. LAVE J, WENGER E. Situated Learning: Legitimate Peripheral Participation. Cambridge: Cambridge University Press. 1991.

17. DREYFUS HL and DREYFUS SE. *Mind over Machine*. Oxford, UK: Blackwell. 1986.

18. OLTEDAL HA. Safety culture and safety management within the Norwegian-controlled shipping industry: state of art, interrelationships, and influencing factors. Stavanger: University of Stavanger, ISBN 978-82-7644-464-3. 2011.

5. AUTHORS BIOGRAPHY

Dr Helle Oltedal is the leader of the MarSafe group at the Western Norway University of Applied Sciences. Her research interests include Human Factors and organisational safety management.

Richard Wild is Senior and Examining Pilot at Harwich Haven Authority, and a PhD student at the Western Norway University of Applied Sciences.

Dr Margareta Lutzhoft is a Professor in the MarSafe group at the Western Norway University of Applied Sciences. Her research interests include Human Factors, human-centred design and the effects of new technology.

EXPLORING DESIGNS OF AUGMENTED REALITY SYSTEMS FOR SHIP BRIDGES IN ARCTIC WATERS

S Frydenberg, K Nordby and **JO Eikenes,** The Oslo School of Architecture and Design, Norway

SUMMARY

It is critical for mariners to maintain awareness of what is happening outside the ship. However, an increasing number of bridge systems force users to switch rapidly between an outside view and the screens inside. Augmented reality (AR) technologies may solve these problems by overlaying the physical world with digital content such as graphics and audio. Designing AR systems is a new and complex design space for interaction designers as it requires an extensive understanding of the users' context and how the technology applies to that context. Understanding the real-world implications of rapidly shifting contextual factors is essential for designing systems that support operators' situational awareness. To investigate which situational premises affect the design of AR systems, we have conducted two early-phase field studies using a broad, multifaceted approach. We present our findings and discuss how the resulting insights may be relevant for building a framework for AR design.

1. INTRODUCTION

Today, mariners on a ship bridge have to exert great efforts to alternate their focus between the separate interfaces on the bridge in order to achieve situational awareness (SA) [1]. At the same time, their perceptual awareness of what is happening outside the ship needs to be maintained, which constitutes a crucial aspect of the demand on the mariners during operations. However, ships' bridge consoles consisting of fragmented and detached systems force mariners to shift their attention (in highly inexpedient ways) between the separate systems interfaces on the inside of the ship bridge and what is happening on the outside.

Emerging technologies, such as augmented reality (AR), may provide new ways of designing user interfaces for ship bridges that meet the unique needs of mariners: to synchronise their personal experience of control with their overall situational awareness, both inside and outside of the vessel [2]. By mixing their perception of the real world with graphical and auditory overlays representing key information, mariners may be able to concentrate their focus on handling a situation to a greater extent. As shown in the aviation industry, a well-functioning mixed reality head-mounted display can have a major impact on workplaces where SA is essential [3,4]. It is likely that AR could function as an expert support system for navigators on ship bridges in the future [5].

Designing AR systems for ship bridges is challenging for interaction designers, as AR introduces new levels of complexity to the design and the use situation. We consider a system complex if, in its entirety, it contains several more diverse features than each individual part contains; this complexity is due to properties such as diversity, interrelations and adaptivity [6]. The dynamic aspects of *environment* and *context* constitute constantly changing key parameters in AR systems. These aspects make the cause-and-effect-relationship of these systems difficult for designers to comprehend – and to account for in the design.

In order to enable the design of future AR-supported systems, there is a need to better understand the possibilities and constraints of designing for AR in a maritime context. To contribute to this area of knowledge, we report on research exploring the human-centred design of AR systems to support SA and decision making for ship bridge crews on Arctic-bound vessels. In this paper, we present our findings and discuss our own experiences from two field studies conducted in Arctic waters to inform the design process of AR. Based on data from the two field studies, the authors' personal experiences and the analysis. These data are implemented to further the design process.

This article reports research that is a part of an EU project, 'Safe Maritime Operations under Extreme Conditions: The Arctic Case' (SEDNA), which seeks to develop an innovative and risk-based approach to safe Arctic navigation, ship design and operations [7].

2. BACKGROUND

Navigators on modern ships work in technologically advanced environments. They need an extensive understanding of bridge systems in addition to navigation skills [2]. Information can be presented in misleading ways in such integrated and networked bridge systems, leading to human errors if the information is not continuously visually compared with the ship's surroundings [2]. In order to correct such errors, it is important for the navigator to frequently check all systems or sensors, such as the radar, to take cross bearings [8] and to maintain a good visual perception of what is happening outside the bridge (through the windows).

Hareide et al. (2018) found that the navigators had insufficient head-up time when navigating. The authors argue that the navigators should address their attention 'most of the time to the outside of the vessel, controlling and comparing the position of the vessel towards the navigation system' [8]. Such problems might be solved by capturing parts of the (or the entire) user interface of the bridge through AR-glasses; then, navigators could, on demand, combine their visual perception with key data from the navigation systems close to the area of interest (AOI) outside the window.

AR is part of a continuum of technologies often called mixed reality [9]. Our work addresses multiple variations of AR technology, as shown in Figure 1. However, our main focus is on AR supported by transparent head-mounted displays, optically superimposing virtual images onto the physical world.

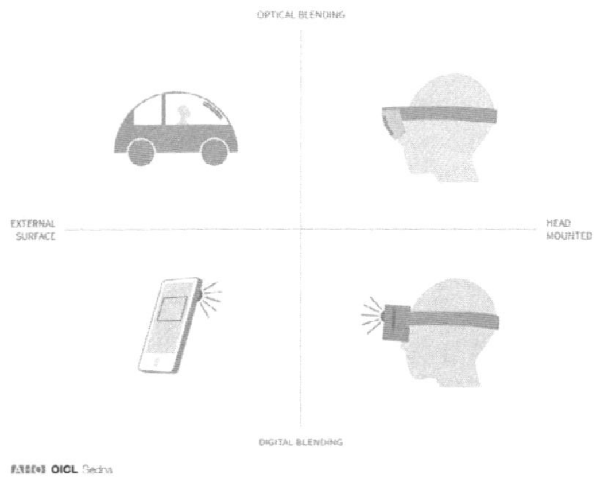

Figure 1: Matrix presenting categorisations of AR as external-surface versus head-mounted and optical versus digital blending.

There has been a steady increase in the number of available head-mounted AR and mixed reality systems on the market in recent years, such as Microsoft HoloLens and Meta 2. Although few of these devices can be realistically implemented to support ship bridge activities in their current versions, they offer good reasons to investigate AR technology as a useful interaction platform for mariners in the near future. Several research projects have looked into how AR can meet ship bridge navigators' need for SA to prevent human error and shipping accidents [5,10–13].

To design interfaces that support a user's SA in safety-critical operating systems is a challenge. According to a study of 100 marine casualties published by the U.S. National Transport Safety Board, poor design of equipment partly caused one third of the accidents [14]. User interfaces designed to integrate with the real world could improve this situation at sea. Improvements may be related to reducing head-down time, aiding the user's ability to directly relate information to their surroundings and superimposing expected ship trajectories onto the physical world. However, aiming at improving a navigator's ability to handle highly variable working situations through design implies designing for *situated interaction* between the navigator and the system. Such a system needs to interpret the user's needs according to the situation they are in, accordingly offering various possibilities for interaction with the system.

Research on collisions suggests that users need visual clues of an impending collision or grounding. A real-world view of the seascape could be augmented using AR goggles, which could work as a feasible method of enhancing SA by reducing head-down time [5]. According to studies from the Naval Academy on the use of AR glasses to provide standard information (e.g. direction, pace, designated distance, bearing, and turn information) in a seascape view during high-speed navigation on littoral waters, eye tracking is a useful method for collecting user requirements [11,12,15].

Designing for AR is not only about designing information according to predefined user requirements but is also about understanding the requirements embedded in a given, yet unpredictable, context. AR interfaces are far more integrated into the user's surroundings than traditional interfaces are. In the development of AR applications for ship bridges, some researchers claim that AR experiments must wait until our understanding of 'what' and 'how' is fully clarified [13]. However, we suggest that an explorative and opportunistic approach will take us further and help us understand the premises of design for AR by iteratively investigating it from many angles. As researchers and designers, we will never fully understand, or be able to map, all situations. Therefore, an important part of our approach is to define scenarios upon which we can base our design decisions. We also propose develop the research in this area as an iterative process of building knowledge, where the interrelated aspects we design for – users, context and technology – are investigated in parallel.

3. APPROACH

The aims of the two field studies were threefold; first, we aimed to investigate the premises and possibilities for designing AR systems to be used by ship bridge crews in different scenarios. Second, we aimed to familiarise ourselves with the context and environment of ship bridges in general and during operations in Arctic conditions, specifically. Third, we aimed to test and evaluate a broad set of methods to identify how interaction designers can approach and define the new and complex design space of AR for ship bridges in the field. These aims were carried out by drawing on the *design-driven field research* [16] model (Figure 2) consisting of three related focus areas – data mapping, experiencing life at sea and design reflection – and the field study processes described in *Implementing Field Research in Ship Design* [17].

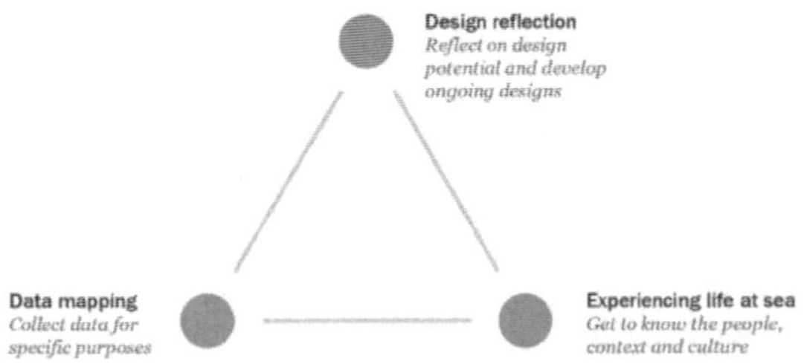

Figure 2: Model for *design-driven field research* developed in the project Ulstein Bridge Concept [16].

The first field study team consisted of a PhD fellow in interaction design, a research assistant and a master's student in industrial design. They conducted the field study for fourteen days on a Norwegian coast guard ship on a research expedition to the West Ice. The research focused on identifying the tasks, the systems used to do these tasks, and the workflows of the bridge crew in selected scenarios. Further, they mapped the physical environment of the bridge and investigated how varying conditions, such as movement and light, affected the needs for, and abilities to handle, system information.

The second field study involved two designer-researchers from the Oslo School of Architecture and Design (AHO) and three researchers from the HALPIN Centre on one of the Swedish Maritime Administration's icebreakers that operates in the Bay of Bothnia. This field study focused on testing current AR equipment and uncovering premises and possibilities for designing AR solutions for different scenarios.

3.1 DATA COLLECTION

We used participatory observation as an approach to familiarise ourselves with the context of the vessel and to gain insight into the working situation [18]. We conducted short informal interviews, participated in the day-to-day life of the crew, observed the work on the bridge and initiated group discussions about user needs and ideas to meet these [18]. In order to better understand the operations, we supplemented our observation of navigation in dense ice on the West Ice expedition by the eye-tracking device Tobii Pro Glasses 2 [19]. Recordings were conducted according to normal watch procedures while manoeuvring in dense ice from the bridge wing console.

In both field studies, the bridge environments were documented and mapped as plan diagrams to display the organisation of working stations and the working flow of the actors between them. The actors' movements during different scenarios have significant implications for AR as the surroundings, the field of view and other possible conditions change.

On the West Ice expedition, comprehensive documentation of all consoles on the bridge was carried out to achieve an understanding of the totality of the bridge systems and the current working situation. The implications for AR rely on existing information displays and the possibilities for using suitable areas to embed AR within existing environments.

3.2 AR CONCEPT TESTING

A number of tests were conducted on both field studies to observe and evaluate how graphical content and its placement in the bridge environment worked in various environmental conditions in terms of light and movement. The overall aim of these tests was to investigate premises for designing AR to be used on a ship's bridge. On the expedition to the West Ice, we brought Microsoft HoloLens AR glasses to test simple mock-ups and applications. We brought Meta 2 AR glasses with us on the field study to the Bay of Bothnia in order to carry out specific graphical tests. Both HoloLens and Meta 2 allow users to see and hear graphics and audio overlaid on top of the physical world. A virtual reality (VR) scene driven by the Samsung Odyssey VR headset that showed an existing bridge design concept was tested by the crew. An iPhone combined with AR markers was used for testing ideas in context. In addition, a portable mini projector and techniques such as paper prototyping and photoshopping were used to simulate AR in a ship bridge environment to explore its different aspects. The AR tests conducted during the field trips were documented in a test log. Several ideas were prototyped on board based on iterative feedback from the crew.

3.3 ANALYSIS

After the field studies, we analysed the data from different categories, carried out a collective analysis and compared our findings during workshops following the methods laid out in [17]. The field studies were documented in two reports that were shared between internal (at AHO) and external project members. The reports were also shared with the crew members on the vessels. External domain experts have verified the content of the documentation in the coast guard vessel report.

The data gathered in the field were used to define and develop specific operational scenarios. We identified and developed the chosen scenarios in consultation with the two ships' bridge crews. After the field studies, the data were analysed and evaluated in team workshops. Further, the scenarios were refined and presented to domain experts as well as the ships' bridge crews from the field studies for verification. We developed a total of ten different scenarios: inshore navigation, offshore navigation, navigation in dense ice, helicopter operation, sea bear (small boat) operation, rescue operation (in which the rescue ship is stuck in ice), escort operation, parking operation, vessel handover and tugboat operation. The aim was to achieve a broader understanding of occurrences in the chosen scenarios, of the actors and their roles, of the bridge functions used and of the communication needs between actors. Scenario mapping is used as a starting point for designing AR concepts.

4. RESULTS

We have divided our main findings of the study into three categories: 1) the technical challenges of using current AR technology on a ship's bridge, 2) premises for the placement and appearance of visual AR and 3) example AR concepts based on scenario observations.

4.1 TECHNICAL CHALLENGES OF USING CURRENT AR TECHNOLOGY ON A BRIDGE

It is important to test AR technology in real conditions in order to better understand the inherent limitations and opportunities of the technology for design purposes. We do expect the technology to improve over time; however, many of the issues we find now, such as contrast and transparency, will most likely persist in future generations of hardware. In the following section, we describe the challenges of using current technologies in maritime conditions.

4.1 (a) Environment Mapping

The current generation of AR headsets has sensors that scan a user's physical environment in order to generate a 3D map. This map is used as a foundation for identifying users' place in 3D space and to allow 3D graphics to be positioned in direct relation to the real world. However, in our tests of Microsoft HoloLens and Meta 2 in the field, we experienced significant tracking problems on the ships' bridges. Both headsets demonstrated problems while mapping both internal and external environments, probably because of the optical difficulties entailed in the environment (i.e. large windows and varying light conditions).

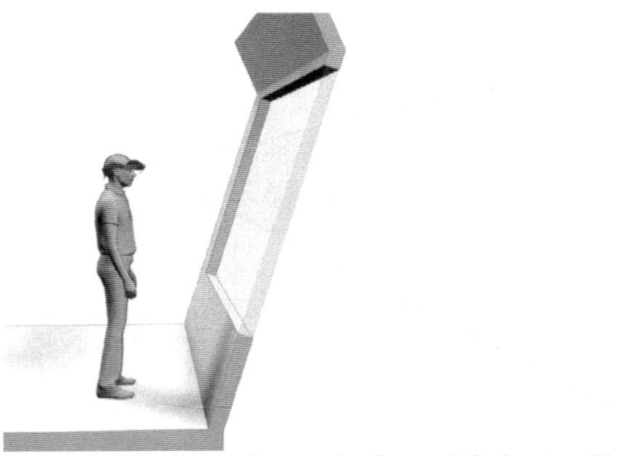

Figure 3: Headsets where only forward-facing tracking cameras are used might not capture any objects on the bridge when the user is standing close to a window, resulting in poor tracking.

The design of good AR systems depends on good tracking in order to fix graphics to internal surfaces, such as panels, windows and consoles, and to pin graphics to external objects, such as other vessels, or to surfaces, such as that of the ocean, that of ice or the horizon. Neither Meta 2 nor the HoloLens managed to offer consistent environmental tracking to support their use in SA scenarios. This is a problem that needs to be solved if these devices are to be used in maritime conditions.

4.1 (b) Contrast and Light Conditions

The differences in ambient light between inside the bridge and outside the vessel are extreme during daytime, especially in ice-filled waters, due to sun reflections. The eye needs time to adjust to the dark areas inside the bridge and the light areas outside. This is a general problem ship bridge crews have to handle, and it is often met by wearing sunglasses or using sun blinds on the windows. During our field study tests, we noted significant visibility problems with the AR devices in terms of displaying graphics onto very light areas (e.g. outside the window during the day) due to the AR devices' combinations of real visual perception and graphics.

At night time, dawn, dusk or in dark weather conditions, current AR devices can display graphics sufficiently on the surface of a window and beyond it without further modifications. In daytime, there is a need for technological improvement for AR glasses that combine real perceptual vision with graphics, such as Microsoft HoloLens and Meta 2.

Differences in light conditions affect how we design AR interfaces. User interface colour schemes need to be adapted so as to offer optimal readability against shifting backgrounds (Figure 4). These problems with light conditions can be avoided using digital blending, where users view the world as a blend of virtual information and digital video of the real world. Such AR solutions can reduce extreme brightness before combining live footage with augmented graphics. Such digital blending is common in mobile phone applications.

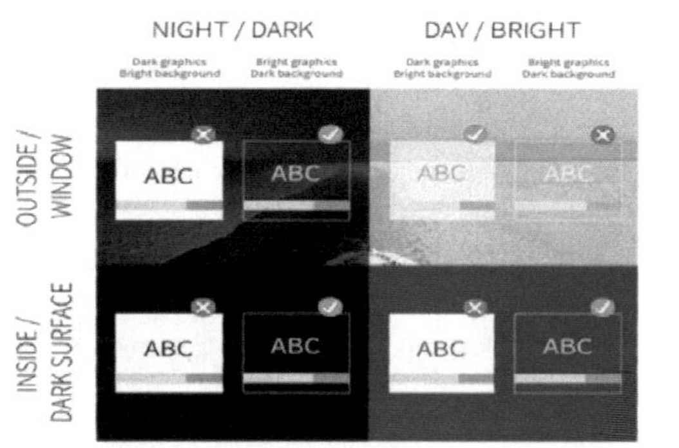

Figure 4: An early schema proposing how colours, fonts and backgrounds should be used in AR in different light conditions without focusing on the visibility problem.

4.1 (c) Movement

Internal movement, which we define as a user's way of moving in their environment, must always be thoroughly mapped in all use cases. Internal movement patterns on ship bridges vary greatly according to the size and design of the ship bridge, as well as the operations and the tasks of the crew. In both field studies, the vessel and operation types required a considerable amount of movement from the navigators on the bridges in most situations. Movements included full body, hip, shoulder and head rotation on a navigator's own body axis. For designing AR, this implies that a navigator needs to use their natural visual perception to a great extent, viewing their environment from many angles to continuously maintain SA; thus, graphics must support this need by careful integration of SA support that does not disturb the user's ability to observe the real world.

External movement, which we define as motions in the user's environment in which the user is located, constitutes an aspect that significantly affects the design of AR for the maritime domain. A ship's bridge is continually moving, and the degree of motion can increase dramatically in demanding weather conditions. External movement forces the crew to focus on keeping their bodies in balance by trying to equalise their weight through internal movement, stay close to handrails and use their visual perception to orientate themselves. This means that the AR system must be able to support a wide range of movement conditions, from calm to demanding. This adds complexity to how we overlay graphics in the real-world view as well as how users will interact with the AR system. There will most likely be a need to design interfaces that can adapt to various use situations, as well as offering users multiple modes of interaction [20].

In the West Ice field study, we experienced a wide range of ship motions, from no waves to storm conditions. We found that using AR while the ship was in heavy motion was difficult using current technologies. We tested the equipment ourselves in different conditions and found it difficult to locate ourselves and balance our bodies while using AR glasses.

The occurrence and degree of nausea is highly individual and might be reduced with experience at sea. However, we argue that AR systems must be designed with these concerns as important test parameters to develop guidelines. Therefore, the combination of external and internal movements must be considered an important area for further research.

4.2 PLACEMENT AND APPEARANCE OF VISUAL AR (GRAPHICS)

AR graphics can be placed in a user's surroundings in endless ways. We define three types of placement used to categorise design concepts: affixed to the body, affixed inside the bridge and affixed outside the bridge. Investigations of placement and appearance of AR graphics in field studies, for all three categories, is highly dependent on the physical bridge design, such as the size and placement of the windows and the layout of the bridge consoles. Since these factors may vary between vessels to a great extent, they should be considered important aspects to examine in the AR design from case to case. However, we suggest that overall guidelines can be developed in the framework for AR design for placement and appearance of AR graphics according to generalisable factors, such as tasks, movement and SA support. To gain an initial understanding of this, we present selected areas for placement and suggest aspects of appearance to be tested further.

4.2 (a) Placement: Fixed to a Body Sphere

AR Graphics can be affixed to a predefined sphere around the user, which is locked to their head, shoulder or hip movement. The size of the sphere defines the distance between the user's eyes and the graphics. The graphics will follow the user's movements, unlike graphics that are fixed to surfaces or points in the environment. Fixing graphics to a body sphere requires no environment tracking, whereas graphics fixed to the inside or outside of a bridge rely on the AR system's tracking of the surroundings to make a functioning projection.

Figure 5: Representation of body spheres for fixing AR content divided into two categories: head-locked content and shoulder-locked content.

In the field studies, most of the premade AR test files we tested on board had graphics affixed to a body sphere. The users experienced annoyance during many of these tests, as the they had to switch their focus between the graphics and the moving environment, rather than looking at graphics embedded in the environment. This was especially evident when the users moved around and when the environment was moving. This experience can be compared with having a stain on one's eye. When the vessel is moving, the body relies strongly on visual perception to maintain balance, which forces the user to focus on the environment, so the stain-effect increases.

Based on these experiences, we suggest that the majority of graphics should be pinned (fixed) to points or surfaces inside the bridge environment or beyond it (to the outside).

4.2 (b) Placement: Fixed to the Inside of the Bridge

Because of glare and the high light contrast on the bridge during daytime, surfaces on the inside of the bridge provide better opportunities for achieving clear graphics than do those on the outside. Many bridges contain areas close to the windows that would be suitable for AR placement, such as panels below and above the windows, the window mullions, and unused surfaces on the consoles. These areas should be considered suitable for overlaying AR graphics containing a larger amount of information since the graphics will not interfere with other important visual information.

Placement on surfaces inside the bridge maintains the existing user situation of switching between information that is relatively close to the eye (1–5 meters) and what is happening outside the vessel, which is relatively far away from the eye (10–∞ meters).

4.2 (c) Placement: Fixed to the Outside of the Bridge

Outside placement of AR graphics could be suitable for a number of AR design concepts that have been evaluated as potentially useful solutions by the bridge crew. Examples are overlaying AR graphics of AIS information pinned to other vessels (Figure 9), projecting key information onto the surface of ice or water during high-intensity operations (Figure 8) or projecting data about ice conditions and leads on the ice, which may support tactical navigation in dense ice (Figure 8).

Placement outside the bridge can be divided into area categories, such as the following:

- On an object (e.g. a vessel, a floating object such as a container or an onshore [or on ice] object such as a polar bear)
- Above the horizon
- On the ocean surface
- On the outside of the vessel (e.g. outside windows, beside the bow or on deck)

Information projected onto the outside of the vessel, close to the windows, is exemplified in Figure 6, where widgets containing key information are displayed outside the windows as the navigator is manoeuvring through dense ice.

Figure 6: Placement in the air outside the window.

By bringing key information closer to perceptual information while looking outside the window, the head-down time of the navigator may be reduced. This implies that if the layout of key information is cleverly designed in order to facilitate a dual information stream, referring to both perceptual and digital information, it could potentially strengthen the navigator's SA.

4.3 DEVELOPMENT OF AR CONCEPTS

As part of the field studies, we developed many concepts on the ship while communicating with the crew. Conducting design conceptualisation in the field is important because it allows us to explore opportunities within the full context, including the participation and comments of the ship's crew [21]. In the following sections, we present two examples of concepts generated at sea that were subsequently visualised in the lab.

4.3 (a) Problem during Navigation in Dense Ice

Navigation in dense ice requires the ship to rupture hard surfaces of ice and manoeuvre between floes. Even though icebreaker vessels can break relatively thick ice, the navigator has to pay close attention when monitoring the ice conditions outside the window in order to manoeuvre safely and efficiently. At the same time, monitoring key information from the bridge systems, such as speed, heading, and machine power, is also necessary. To constantly switch focus between several AOIs inside the bridge and the ice outside makes it difficult for the navigator to maintain good SA. This is due to several factors, including the differing focus distances and light conditions between inside and outside, the cognitive workload required to align and compare several types of data, and the constant need to interpret one's visual perception.

In our field studies, the navigators found it difficult to fully understand the implications of important navigational information, such as radar images of ice or *no-go zones*, from the bridge systems in relation to the real world outside. They expressed a need to better connect information from applications, such as Electronic Chart Display and Information System (ECDIS) and radar, with the outside view.

Based on this need, we present a design concept in Figure 7 where geographical and navigational information is presented as a transparent layer on top of the ocean surface. The figure indicates the no-go zones, the heading and the estimated path of the ship.

Figure 7: Design sketch with graphics placed on top of a photo, illustrating what the AR view could look like when geographical and navigational information is overlaid on the ocean surface.

We discussed early sketches of this concept with both crews during the field studies, and they considered this concept useful for tactical planning of routes and navigating in closed waters, especially during conditions with poor visibility, such as due to darkness or fog. This concept was originally suggested by the navigators in one of the field studies as a way to help less experienced members of the ship's bridge crew sail in Arctic waters, as this design can supply the user with information about ice conditions, for instance, by implementing information from drone photos into the AR overlay projected onto the real world, such as on the ice surface.

A solution to address other needs in this scenario is given in Figure 8. This solution involves placing small information widgets (e.g. for machine power or heading information) close to the navigator's AOIs near the bow of the ship.

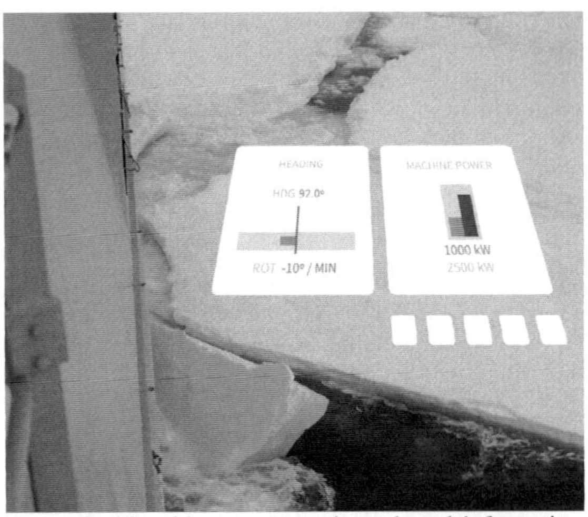

Figure 8: AR widgets representing selected information from the Conning system placed onto the ice.

The concept shown in Figure 8 was informed by the eye-tracking data we collected, showing that users switched rapidly between looking at key information on the inside and at the ice outside. The feedback from the crew was positive regarding earlier sketches of this concept as it may reduce the problem of constant alteration of one's line of sight and mental focus. It might also make comparison of data easier for the navigator, as the information can be represented and placed in a more suitable location.

4.3 (b) Problem during Escort Operation

This scenario describes an on-duty icebreaker that escorts another ship. The ice breaker creates a channel in the ice through which the other vessel can follow. This scenario presents challenges for navigators, such as monitoring the speed and distance between the ships.

In Figure 9, we present a design concept that meets the needs of the navigator by connecting labels or information feeds to the icebreaker vessel, enabling the connection of the real-world view with key information needed in this specific situation.

Figure 9: Design concept made by placing graphics on top of an image to show how key information can be connected to another object.

We showed early sketches of this concept (Figure 9) to the crew in one of the field studies. They considered the idea of connecting information labels to objects in the external environment as a useful one. Especially during operations that demand a certain degree of control over other vessels, such as tugboat, escort and sea bear operations, this virtual labelling could support the navigators' SA and allow them to achieve a better overview of the situation.

In other conditions it might be beneficial to have different representations of the same type of information, for example during complete darkness, or if the view is obstructed by large objects. In the design concept in Figure 10, key information is displayed on one of the dark panels below the window in order to keep the information close to the vessel(s) outside (which now can only be perceptually identified by lanterns). In addition, the icebreaker vessel's own key information is placed next to the escorted ship's information to make comparisons easy.

Figure 10: Design concept to indicate a ship's own relation to other ships by displaying information on a dark panel below the window. Prototyped by projecting graphics onto surfaces on the bridge using a portable mini-projector.

The development of scenarios was central to our concept development. The ten scenarios provided us with a better spatial and temporal understanding of selected situations where conditions, circumstances and needs were defined. The scenarios enabled us to structure ideas and develop concepts according to specific frames. Concepts staged within a scenario can, in some cases, be easier to communicate than abstract or general concepts because both experts and users are familiar with the scenario, which defines clearer parameters from which to evaluate the concept.

5. DISCUSSION

The two field studies have given us a wider understanding of the design problem; we collected many types of data during the studies that we have built upon in the further design process. We have presented the results after analysing these data and will discuss their implications for further research.

The overall results of the two field trips have shown that AR has the potential to be used as a rich interaction medium for navigators on the bridge of a ship. Through these field studies, we found that AR can be used to display a broad range of information necessary when operating a bridge. We suggest that AR has the potential to be a support system that can be combined with a physical bridge or even fully integrated with a complete bridge operation system.

Our research indicates that AR has the potential to convey many types of information. Research on AR in the maritime domain has focused on SA support [3,4]. However, our research is focused on the *interaction design* of AR. Our results point to a broader range of AR concepts than have been described in previous research. We suggest that a number of applications, possibly the whole integrated bridge system, may be mediated through AR glasses. Our exploration of new ways to present key information according to specific scenarios demonstrate that AR has great potential for presenting information in new and more situated ways.

The scenarios we have developed enabled us to structure design concepts and demonstrators according to realistic situations and defined needs. The selected scenarios also strengthen the reliability of communications with outside actors. Possible disadvantages of focusing on selected scenarios include the lack of total needs among the crew members in all situations. It might be problematic to transfer abstract concepts that have been developed for specific scenarios. Finally, the selected scenarios could be too narrowly defined as it might not be possible to capture all conditions in a given scenario that could affect the users' needs. This should especially be considered when working with a ship's bridge crew as a use case, as it is a context exposed to highly varying conditions.

We have described some of the technical challenges of using current AR technology on a ship's bridge, including environmental mapping, movement, and contrasts and light conditions. Our findings clearly show that the immaturity of the current technology and equipment causes extensive problems when these are used on a ship's bridge. There were problems in testing all prototypes, which partly led the research to focus on the aspects that could be tested. However, we have tried to look past current technical challenges of today, and rather used design techniques such as manipulation and staging for sketching and testing ideas. We suggest that the user's needs should be considered to understand how the technology ought to work, and this can be defined through a communicable design concept.

Our work suggests that designing AR systems for ship bridges is a complex design problem that needs to be approached iteratively from multiple angles. Utilising a range of methods during field studies has proven useful for investigating the conditions for AR described in this paper, which can be used in building a conceptual framework for AR. We further suggest that common guidelines are needed for user interfaces and interactions with all bridge applications in order to move towards an integrated AR bridge system.

6. CONCLUSION

In this paper, we presented findings from two field studies conducted on icebreakers with the aim of exploring premises and possibilities for designing AR systems for Arctic-bound ship bridges.

Our study shows that designing AR applications for the maritime domain involves specific challenges, such as ship motion, light conditions and a large variation in user movements and tasks. Further, we found that the current state-of-the-art AR hardware does not function optionally in a ship bridge environment with its variable light and large reflective windowpanes. However, we also found many opportunities for supporting ship bridge crews' work using AR technology.

Based on our experiences, we argue that designing AR for ship bridges is challenging; however, if the technical challenges can be solved, AR may improve maritime operations in the future.

7. ACKNOWLEDGEMENTS

The research presented in this paper was funded by the EU project SEDNA. Thanks to Nicholas Stevens for contributing with ideas, observations and technical skills. Thanks to the SEDNA team for their participation in the field studies, their input and their contributions to furthering the design processes. Thanks to Odd Sveinung Hareide and Robert Lynch for organising the field studies, and Hareide for an expert evaluation of one field study report. Thanks to NTNU Ålesund for lending eye-tracking equipment. Special thanks to the Norwegian Coastguard, the KV Svalbard and crew, The Swedish Maritime Administration, the Icebreaker Atle and crew.

8. REFERENCES

1. LÜTZHÖFT, M., NYCE, J., 'Integration Work on the Ship's Bridge', *Journal of Maritime Research*, 2008.
2. HAREIDE, O.S., OSTNES, R., MJELDE, F.V., 'Understanding the Eye of the Navigator', *European Navigation Conference Proceedings*, 2016.
3. MELZER, J., 'Head-Mounted Displays', In: SPITZER, C., FERRELL, U., FERRELL, T., editors, *Digital Avionics Handbook,* 3rd edition, Taylor & Francis Ltd, 2017.
4. RASH, C. E., 'The Potential of an Interactive HMD Helmet-Mounted Display' US Army Aeromedical Research Laboratory, 2009.
5. BALDAUF, M., PROCEE, S., 'Augmented Reality in Ships Bridge Operation', *ISIS 2014 – International Symposium Information on Ships*, 2014.
6. LURÅS, S., LÜTZHÖFT, M., SEVALDSON, B., 'Meeting the Complex and Unfamiliar: Lessons from Design in the Offshore Industry', *International Journal of Design*, 2015.
7. SEDNA-project.eu [cited 2017 Nov 28]. Available from: https://sedna-project.eu/.
8. HAREIDE, O.S., 'Improving Passage Information Management for the Modern Navigator', *IALA 2018 Korea*, 2018.
9. SILVA, A., SUTKO, D.M., *Digital Cityscapes: Merging Digital and Urban Playspaces*, 1st printing edition, New York: Peter Lang Inc., International Academic Publishers, 2009.
10. BENEDICT, K., KIRCHHOFF, M., GLUCH, M., FISCHER, S., SCHAUB, M., BALDAUF, M., 'Simulation-Augmented Methods for Safe and Efficient Manoeuvres in Harbour Areas', *International Journal on Marine Navigation and Safety of Sea Transportation*, 2016.
11. HAREIDE, O.S., MJELDE, F.V., GLOMSVOLL, O., OSTNES, R., 'Developing a High-Speed Craft Route Monitor Window', In: *Augmented Cognition Enhancing Cognition and Behavior in Complex Human Environments*, Cham: Springer, 2017
12. HAREIDE, O.S., OSTNES, R., 'Scan Pattern for the Maritime Navigator', *International Journal on Marine Navigation and Safety of Sea Transportation*, 2017.
13. PROCEE, S., BORST, C., VAN PAASSEN, M.M., MULDER, M., 'Toward Functional Augmented Reality in Marine Navigation: A Cognitive Work Analysis', *16th Conference on Computer and IT Applications in the Maritime Industries*, 2017.
14. ROWLEY, I., WILLIAMS, R., BARNETT, M., PECKAN, C., GATFIELD, D., NORTHCOTT, L., et al. *Development of Guidance for the Mitigation of Human Error in Automated Ship-Borne Maritime Systems*, Swansea: Maritime and Coastguard Agency, 2006.
15. HAREIDE, O.S., OSTNES, R., 'Maritime Usability Study by Analysing Eye Tracking Data', *The Journal of Navigation*, 2017.
16. LURÅS, S., NORDBY, K., 'Field Studies Informing Ship's Bridge Design at the Ocean Industries Concept Lab', *Human Factors in Ship Design and Operation: 26–27 February 2014,* London, UK: Royal Institution of Naval Architects, 2014.
17. GERNEZ, E., NORBY, K., 'Implementing Field Research in Ship Design', *In Press.*
18. DEWALT, K.M., DEWALT, B.R., *Participant Observation: A Guide for Fieldworkers*, Rowman Altamira, 2011.
19. Tobii Pro Glasses 2 Wearable Eye Tracker 2015 [cited 2018 May 14]. Available from: https://www.tobiipro.com/product-listing/tobii-pro-glasses-2/.
20. NORDBY, K., LURÅS, S., 'Multimodal Interaction for Marine Workplaces Used as Strategy to Limit Effect of Situational Impairment in Demanding Maritime Operations', *International Conference on Marine Design 2015, 2–3 September 2015,* London, UK, 2015.
21. LURÅS, S., 'Layered Scenario Mapping: A Multidimensional Mapping Technique for Collaborative Design', *CoDesign*, 2016.

9. AUTHORS BIOGRAPHY

Synne Frydenberg has been a PhD fellow since 2017 at The Oslo School of Architecture and Design. She also works part time as a design researcher within the SEDNA project. Her research interests lie in designing for situated interaction in complex professional settings, such as a ship's bridge. Before starting her PhD, she worked six years as an interaction designer in the Norwegian design consultancy Making Waves and in the digital department of a Norwegian publishing house, Gyldendal.

Kjetil Nordby (PhD) is an associate professor at The Oslo School of Architecture and Design. He is a research manager for Ocean Industries Concept Lab and the project leader of the OpenBridge project. His previous experience includes interaction design, industrial design, design management and research management within ocean industries.

Jon Olav Eikenes (PhD) is a designer and researcher at The Oslo School of Architecture and Design. He works part time as a design researcher within the SEDNA project. His previous experience includes research on visualisation and animation in interfaces. He has worked several years as a designer and contributed to the design of several products for ship bridges.

PRESENTATION

VIRTUAL REALITY FOR HUMAN FACTORS ENGINEERING

K. Goh, Knud E Hansen & **RJ Spencer**, Stirling Labs

1 INTRODUCTION

The ship building industry has always been at the forefront of using Computer Aided Design (CAD) tools, for example mathematical surface modelling for the complex geometry of ship hulls. Today an entire ship complete with all internal details can be created in a three-dimensional Virtual Reality Environment (VRE). In this paper the authors describe some of the current issues facing designers regarding Human Factor Engineering (HFE) in ship design and explore the use of the ShipSpace™ Virtual Reality collaboration toolset and how the tools are being further developed and tested for this purpose.

2 BACKGROUND

2.1 VIRTUAL PROTOTYPING

Virtual Prototyping (VP) is the use of a virtual environment for design and evaluation of the design. It is replacing more traditional methods of prototyping such as scale models and mock-ups. There has been considerable research into using Virtual Reality Environments (VRE) for training and education. However, the use of VRE for prototyping has not been given as much focus.

It is important to make the distinction between what constitutes a VRE. While a virtual environment can be considered as any visual method of presenting a digital 3D construct, such as objects designed on a CAD application with a computer monitor. A VRE requires a much higher level of immersion, such that the observer feels like they have presence in the digital environment.

Previously the authors have discussed [1] the varying ability of people to visualise a ship design by reading drawings that show views, elevations, and sections. Goh and Spencer go on to describe the limitations of using CAD systems on a screen display for understanding complex VREs like a ship and demonstrate why the recent advent and application of consumer VR hardware is such a powerful visualisation and communication medium for design and engineering.

2.2 WHAT IS VIRTUAL REALITY

Virtual Reality provides a computer generated visual experience that tricks your mind into reacting as if you are in a real physical environment, even though you are just seeing a digital virtual construct. Key aspects of achieving this immersion include:

Immersive vision – Wide field-of-view such that you do not feel you are looking through a window or frame.

- Stereoscopic vision - You can use two eyes to gauge size and distance as you do in real life.
- Vision tracking - The view is coordinated with your head movement without noticeable lag.
- Graphical stability - The visual display is smooth, without jerkiness, stutter or drop-out. Real world concepts – Notion of up and down is preserved, simulated gravity constraining you onto a surface., etc.
- Interactive - You can interact with the virtual environment.

2.3 IMPLEMENTING VIRTUAL REALITY

Virtual Reality is best achieved with a Head-Mounted Display (HMD), highly accurate positional tracking system and computer software that updates the display as the user moves with sufficient fidelity that the user can sustain the illusion that they are immersed in the virtual environment displayed.

Projection caves and domes are sometimes said to provide a VR-like experience, but they have many disadvantages compared to HMDs [1]. In particular objects in the VRE that are close to the observer cannot be displayed or approached as they are removed from the image projection when they 'enter' the cave or dome. This make them particularly unsuitable for HFE which is primarily interested in direct human interaction with the environment. In addition, the view in a cave or dome is fixed and does not change with the observer. Therefore, the perspective is only correct in one singular position of the user's head. If the user only rotates their head the perspective may be correct, while if the user translates their head by leaning from one side to the other or squatting down, the view the observer sees will no longer be accurate.

By comparison, the high precision tracking of HMDs allows the observer to study objects in the VRE up close and peer around or under objects. With a HMD the users perspective of the VRE will always be correct regardless of where the objects are or where the user is looking. Therefore, the authors have only considered HMD implementations of VRE for this study on HFE.

2.3 (a) Benefits of ShipSpace VRE for HFE:

- Most realistic way to experience an unbuilt design without using any physical constructs
- Realistic sense of scale and arrangement
- Sight lines can be checked
- Work areas can be checked
- People can collaborate on the design together even though they may be miles or continents apart
- No need to travel to discuss leads to more frequent communication on design issues
- Easier and immersive access to the design means experts can more easily add value to the design process

2.3 (b) Limitations of VRE for HFE

- Requires a 3D model, which can be expensive if it is not already being created as part of the design process
- Realistic tactile interaction with the virtual environment is limited
- Navigation is unrealistic (teleporting)
- Poor implementation can cause motion nausea and migraine
- Some people don't feel comfortable using HMDs for reasons of vanity or having previously had unpleasant experiences with poor implementations

2.4 THE SHIPSPACE™ VR TOOL

ShipSpace™ is a VR design visualisation and collaboration tool developed by Stirling Labs for use in the maritime and naval design, engineering and construction process. The system reads normal CAD files directly and displays the 3D VE on a HMD, complete with all of the detail at very high fidelity. Special techniques are used to maintain rendering latency of less than 11ms "motion to photon" so that immersion is maintained without discomfort, and surfaces and light are treated carefully to create a strong sensation of presence. ShipSpace provides a range of sophisticated tools to interrogate the model in various ways, such as displaying metadata, "x-ray vision", controlling object layer visibility, measuring distances and many others. The tool enables up to 64 users to collaborate simultaneously in the same VRE.

The authors have tested the ShipSpace™ tool for realism, accuracy and the ability of people to learn how navigate and utilise simple tools in the VRE which consisted of a number of different ship models. The findings indicate that the tool provides a highly convincing visualisation medium that enables all people, regardless of spatial skills, the ability to understand design and collaborate more completely. The authors propose that this is primarily due to the immersiveness of the VRE and the more natural human interface, specifically how the users view is coordinated with the position of their head.

2.5 VR APPLICATION TO HFE OF SHIPS

There are many areas of a vessel that can benefit from an immersive and realistic design VRE. Arrangement of areas such as the bridge, galley, hospital and mooring stations need to be highly optimised for effective operation often with many team members. With the ability to have multiple users in the same VRE, ShipSpace™ enables teamwork to also be tested.

Machinery spaces such as engine rooms, funnel casings, A/C stations and thruster rooms, are congested spaces with multiple pieces of equipment and a confluence of piping, ducting and cabling snaking throughout. Such equipment must be accessible for regular maintenance and removal if necessary. When designing for maintainability, it is often necessary to seek the advice of the equipment manufacturer. The ShipSpace™ tool enables designers and experts to more easily collaborate together on the space requirements regardless of where they may be residing in the world. The VRE tool provides a highly realistic environment to discuss the details of the design.

For the purposes of this paper the authors have chosen to focus on the design of the bridge and engine room.

3 HFE WITH VRE FOR BRIDGE DESIGN

3.1 BRIDGE MOCK-UPS

The bridge of a vessel is typically the centre of operations for navigation, communication and planning. Having an optimised layout is crucial for effective operation of the vessel. Not only must the bridge control stations and work areas be arranged correctly, but also the sight-lines from the control stations through the windows must meet various rules and requirements. Often a full-scale mock-up of the bridge is built to verify the layout design of the bridge. Faux consoles will be constructed, complete with paper prints of instrument panels, displays and controls stuck on. Masters and navigators can then role play various vessel operations and scenarios to test the bridge layout. Any modifications required can be applied before committing to the actual bridge construction.

3.2 ANALOGUES AND ACCURACY

In her review of virtual environments, Stanney [1] states virtual experiences for training cannot fully replace the benefits of real experiences. She goes on to state that there is not yet a full understanding on how best to use the technology, and the technology does not always meet the expectations of its users due to issues such as cyber sickness and lack of presence. Popescu [2] had previously noted that the number of sensory modalities stimulated, and the quality of this multi-sensory interaction are critical to the immersiveness and potential effectiveness of virtual environment systems.

The authors suggest that a VRE does not need multi-modal stimulation to be beneficial for the purposes of HFE. This is compared to training where the idealised goal is to make the training experience indistinguishable from reality. Analysis or validation of the design will be undertaken by experienced personnel who will role-play only to the extent necessary to prove the design without needing necessarily to play out every step. Furthermore, experienced personnel, unlike trainees, will already be very familiar with the look and feel of the various bridge control stations and can use memory and imagination to compensate for any deficiency in the realism of a VRE.

On the other hand, from the design and engineering perspective, it is not necessary to make the VRE realistic other than from pure geometric and spatial accuracy. It is normal for the VRE of ship design to be coloured unrealistically, but instead according to an identification scheme: steel structure in brown, cooling pipes in blue, fuel pipes in yellow, fire main pipes in red etc. When assessing a design with respect to HFE, it would be preferable that no augmentation of such faux coloured VREs is necessary, while maintaining its effectiveness for HFE.

Analogue / Accuracy	Geometry	Visual	Physical	Tactile	Sight
Training	•	•	•	•	?
HFE Mock-up	•	?	•	?	
HFE Virtual Prototype	•	?			•
Detail Design	•				
Concept Design		•			

Table 1: Accuracy of Analogues during the design process

Table 1 shows the accuracy of the analogues at the various stages of design and purpose. Geometric accuracy refers to the analogue having the correct dimensions. Visual accuracy refers to the analogue looking like the finished vessel. Physical accuracy refers to the analogue having a correct physical environment. Tactile accuracy refers to finer details of the Human Machine Interface (HMI) such as buttons, switches, controls, handles and instruments. Sight accuracy refers to the ability to check sight-lines.

The reader will note that the 'Sight' accuracy of 'Training' is in question, when this would be expected to be of high important. Often, simulators used for training use a mixed reality approach with a realistic physical environment and screen or projector displays. With such a setup, the accuracy of the sight-lines is often inaccurate since the position of the user's eye can move while the view shown through the windows will not change. Although an HMD can show sight-lines with high accuracy, the need to see and touch the physical environment of a simulator means that sight-lines are compromised by using screen displays. To overcome these issues, there has been some experimentation of combining video images of the real environment with computer generated virtual environment onto an HMD, such as the Rockwell Collins Coalescence system [3]. However, the quality and stability of the merge environments still needs to be improved. One of the issues is that the video cameras capturing the view of the real environment are located 200mm in front of the user's eyes, and thus give a distorted view.

3.3 TESTING OF MOCK-UPS & VIRTUAL REALITY FOR H.F.E.

Note that the requirements of virtual prototypes for HFE will fall somewhere between that of design and training. To what extent VPs requires visual and physical accuracy is being tested using the ShipSpace™ VRE system. The concept of presence has already been tested with the ShipSpace™ VR tool [1] using the moonpool test. The large majority of subjects were not able to walk into the moonpool for fear of falling, despite consciously knowing that they were in a safe office environment.

The current test concerns the design of a bridge wing station operated during the berthing a ship. The bridge wing typically has many windows, so the operator can see forward, aft along the wharf and also directly down to the wharf. The test subjects are randomly assigned to 3 groups, one group looks at a plan drawing of the bridge and are asked to visualise what they think the bridge wing will look like and then evaluate the design using a questionnaire. The second group enter a physical mock-up of the bridge wing and evaluate the design using the same questionnaire. The final group view the wing station in a VRE using the ShipSpace™ tool and are asked to evaluate the design again using the questionnaire. The study is still on-going due to the difficulty in enlisting sufficient subjects for a meaningful result, as the test requires navigators experienced at handling a ship during berthing.

4 HFE WITH VR FOR ENGINE ROOM DESIGN

Typically, engine rooms are highly congested with machinery equipment and connecting pipes and cables all competing for space. In such spaces there are five main concerns:

- Personnel and tools accessibility to equipment for maintenance purposes,
- Removal space and routes for equipment replacement,
- Minimising piping and cabling complexity,
- Resolving clashes,
- Build-ability

All five issues are inter-related to the extent that finding the optimal solution involves satisfying all requirements, which sometimes conflict and may require some degree of compromise.

4.1 ACCESSIBILITY & MAINTAINABILITY

The normal way of ensuring sufficient maintenance access and part removal paths is to simply follow manufacturers recommendations. However, these recommendations are often conservative and may be inaccurate or incomplete for the sake of simplicity. While this may not be of concern because more than sufficient space has been allowed, there are many vessels such as high-speed craft, warships and submarines, where space is of a premium and optimising machinery arrangements is necessary to achieve a functioning and maintainable design.

The traditional approach to optimising an arrangement is to build a full-size mock-up. However, this is impractical for machinery spaces for a number of reasons. Firstly, it is very difficult to get hold of the actual equipment for the mock-up. The design needs to be completed many months or even years in advance of the construction and shipbuilding economics demand that equipment such as an engine is manufactured and delivered just-in-time to be installed in the vessel. Engines are also long lead time items and it is highly unlikely that they could be delivered early enough for the design phase in any case.

The maker is unlikely to have a spare or sample engine laying around that could be loaned for the purposes of a mock-up. The engine is also likely to have a slightly different configuration based on the required engine specification or options chosen. Even if such a sample did exist, the cost of shipping and returning the sample, often many thousands of kilometres, would be substantial.

While building a mock-up of the engine is possible, this is an expensive undertaking requiring skilled model makers. Models may be simplified to save on time and cost however they run the risk of not capturing the true volume or complexity of the engine due to missing details or over representing the volume by erring on the safe side. By missing or simplifying details there is also the possibility that the maintenance and access trials may inaccurate.

Full size mock-ups take up considerable space and introduce the issue of storage or disposal of the mock-up.

4.2 CLASH DETECTION & BUILD-ABILITY

Typically, during the detail design process, different designers are involved with developing the piping and cabling routes from schematic or single-line drawings. A preliminary space reserve or 'space grab' is often modelled by experienced designers as a first step before detail designers follow or trace over with the final pipe spool design complete with joining flanges, valves, filters and other elements that could be in the pipeline.

Despite the initial space reserve planning, invariably clashes (interference between two or more objects that could not exist in the real world), occur in the detail design process that need to be resolved or de-conflicted. Furthermore, even if there are no clashes, the design must be checked for build-ability by experienced production personnel and possibly modified to ensure that it is possible to construct and install. Build-ability is an area of HFE that is often overlooked, since it does not pertain directly to the operation of the vessel.

The current method of checking for clashes and build-ability is to use a CAD system or 3D model viewing application on a screen display. Most CAD software has built-in clash detection;

however, such checking can often be flawed because many 'correct' instances are flagged as clashes. There is also no capacity to check for build-ability, such as the situation where there is no clash, but insufficient space has been allowed for assembly and disassembly. The authors have experienced on many occasions finding significant clashes while examining a model using the ShipSpace™ VR tool and having great difficulty finding the same clashes using a traditional CAD system and a screen display.

The authors suggest that a HMD is much better than a screen simply because users are able to take in so much more of the environment at any given time. When using a CAD system and screen display the field-of-view (FOV) is limited to 30-45 degrees horizontally. A consumer VR HMD has a FOV up to 110 degrees. This amounts to 5 to 10 times more viewable area when considering a 2:1 horizontal to vertical FOV ratio.

The natural way in which the HMD user changes their view point, by moving their head and body like in the real world, seems to allow much faster and effective interrogation of the digital environment compared to the typical hand operated mouse interface of a CAD and screen display system.

4.3 PROTOTYPING WITH VIRTUAL HUMANS & AVATARS

Due to the impracticalities of building full-size mock-ups, scale models of engine rooms have often been built to test that everything can fit, however this had limited use for HFE. Recently virtual prototyping using a CAD system with an anthropometric human digital model, a so-called Virtual Human, presented in the virtual environment has become the normal method to check for accessibility in tight machinery spaces. There has been considerable research into using virtual humans for testing virtual prototypes. Duffy [5] provides a comprehensive review of virtual human development and applications.

However, using virtual humans for HFE is often a tedious and time-consuming task. Many positions of the virtual human need to be trialled for a comprehensive analysis. Viewing of the virtual human in the 3D model can be difficult because of the congested nature of the machinery space and clumsy manipulation of the viewing system on a screen display. The ShipSpace™VR tool is able to spawn virtual humans that can be posed to check and illustrate accessibility issues. However, the authors contend that directly using the virtual avatar instead is even faster and easier.

There is a fundamental difference between using a virtual human compared to a virtual avatar for virtual prototyping. With a virtual human, the biomechanics must be an integral part of the virtual human model in order for the limits of body and limb movement to be realised when assessing the virtual prototype. With a virtual avatar in VR, the limits of human movement are naturally provided by the user.

With virtual humans the difficulty is not only accurately modelling the biomechanic limits, but also in determining and interpreting what represents natural human movement. Just because a body and limbs can be placed in a certain configuration, does not mean that such a body position is comfortable or even amenable for the particular task at hand. The advantage of using VR and a digital avatar is the user can feel what positions come naturally. The challenge is then capturing the motion of the user and producing an accurate digital avatar within the VRE so that the interaction between the user and objects in the VRE can be accurately assessed.

5 SHIPSPACE™ FUTURE DEVELOPMENT FOR HFE

5.1 ANTHROPOMETRIC AVATAR

There are multiple benefits for HFE by enabling the user to become a more 'physical' entity within the virtual environment by presenting them as a digital avatar. For example, an avatar of the user is necessary for better communication between people in the VRE. Such an avatar could be anything from a cartoon type character to a digital model that closely resembles the user. The concept of embodiment has been tested with the ShipSpace™ VR tool [1]. In the test, the instructor appeared as an avatar in the virtual environment. The avatar was a translucent head and bust wearing a similar HMD to the subject. Even with this simple representation, 100% of subjects accepted the avatar as an embodiment of the instructor.

Checking for clashes and build-ability are possible with a VRE where the user is a ghost in the digital environment. The user feels like they are immersed in another world, however if they look towards their physical body, they will see nothing. They are only a sentient presence in the virtual environment.

It is obvious that a more physically accurate avatar is able to interact more accurately with the virtual environment. For the case where the VRE is being used to study personnel and tool access to various equipment in a congested engine room, it would be very useful to have a digital avatar that moves accurately with the user. It would then be possible to detect or observe from a third-person, clashes between the avatar and the VRE and give better feedback on maintenance and accessibility problems. Furthermore, maintenance tools can also be modelled and wielded by using the standard VR hand controllers as a control proxy to make the maintenance testing more accurate.

There are a number of human movement motion capture methods that can be used for the purposes of accurately modelling an avatar in a VRE. These are mature techniques that have been used in the video gaming and movie industry for decades. However, they are invasive and require the VRE user to wear a full body suit covered with reflector dots. It would be preferable if such suits were not needed. The technique also requires additional video cameras to be set up around the user and a computer to process the images and 'capture' the motion of the user.

The authors are currently undertaking testing using the laser tracking system which captures the position of the HMD and hand controllers. This system attempts to minimise the invasiveness by using fewer tracking points that are more simply worn. The user wears special tracked overshoes to enable foot position to be tracked and also clips a tracking unit on the back of the waistline as a body reference. The head is tracked via the HMD and the hands via the controllers. This is expected to be less invasive than wearing a full body suit. Using a simple anthropometric model of the human body, this allows the ShipSpace™ VR system to create a digital avatar of the user's whole body.

The most complex aspect of human biomechanics is the movement of the arms afforded by the shoulder joint, which can rotate in every axis. Testing is ongoing to determine how best to precisely measure or predict arm position. Solutions being tested are additional tracking points attached to the users' upper arms and combining a computational model.

Other methods of body tracking investigated by the authors involve using photogrammetry. These would not require the user to wear any tracking equipment or special suits, however it would require additional equipment to be installed at the VR station.

5.2 COLLISION DETECTION

As noted above, the purpose of capturing the motion of the user is to create an accurate avatar which can be used to check for interference or collisions with objects in the VRE. In a complete VRE, there is obviously nothing physical preventing the user moving their body into virtual objects. Most of the motion capture techniques discussed only capture a few specific points on the user's body. It is then necessary to model the volume of the user's avatar in a realistic manner, without which collision detection with objects in the VRE is not possible. Since the purpose of such modelling is to test for accessibility and maintainability, it is not necessary to model the user's actual body accurately. Instead, it is more valuable to use a standardised human anthropometric data set, such as the Civilian American and European Surface Anthropometry Resource (CAESAR) [6]. The ShipSpace™ tool is able to rescale the actual tracked position of the user to match the chosen anthropometric data, although reliably apprising the user of this volume is not yet fully resolved. As this method is refined, it will be possible for any user to test accessibility with avatars that are sized according to specific requirements.

With a useful method for modelling the user's avatar established, attention can now be drawn to the issue of collision detection between the avatar and objects in the VRE. Collision volumes for the avatars can be modelled based on the anthropometric data. Collision volumes will also need to be modelled for the objects of interest in the VRE. Interference between the collision volumes of the avatar and objects can then be checked and haptic feedback provided. It is useful for the user to know immediately when they are colliding with an object when checking for accessibility, especially in tight spaces. Some tolerance of the collision detection can be useful to reduce reporting of insignificant collisions, so progressive feedback is provided. A human has a soft body that can be reshaped to partly conform when forced onto objects. Maintenance and removal of parts can also be checked by modelling their collision volumes and enabling users to manipulate them in tight spaces.

6 CONCLUSION

Review into the requirements of HFE for ships bridges and engine rooms shows that virtual prototypes when implemented in virtual reality environments have many advantages over traditional methods such as building mock-ups or using CAD based virtual humans, especially for design collaboration over distance. Initial testing with the ShipSpace™ system for bridge wing prototyping shows the VR tool is highly effective in providing a realistic environment and has the important advantage that sight-lines can accurately assessed. The ShipSpace™ system has also been used for checking for and resolving clashes, plus assessing access and maintainability in ship machinery spaces by using virtual humans. Although effective, further analysis of digital avatars is proposed to enable faster and more natural checking since the user's body motions are captured directly. The ability to scale the motions and apply them to a range of anthropometrically defined avatars is expected to be highly useful for evaluating areas for the range of persons expected to operate and maintain the vessel.

7 BIBLIOGRAPHY

8

[1] K. Goh and R. J. Spencer, "Virtual Reality for Design of New Warship Concepts," in *RINA Warship 2018: Future Surface Vessels*, London, 2018.

[2] K. M. Stanney and J. V. Cohn, "Virtual Environments," in *Handbook of Human Factors and Ergonomics*, John Wiley & Sons, 2012 , pp. 1031-1056.

[3] G. V. Popescu, G. C. Burdea and H. Trefft, "Multimodal Interaction Modeling," in *Handbook of Virtual Environments: Design, Implementation, and Applications*, Lawrence Erlbaum Associates, 2002, pp. 435-454.

[4] Rockwell Collins, "Coalescence™ mixed reality system," 1 10 2017. [Online]. Available: https://www.rockwellcollins.com/Products-and-Services/Defense/Simulation-and-Training/Training-Systems/Coalescence.aspx. [Accessed 25 9 2018].

[5] V. G. Duffy, "Human Digital Modelling in Design," in *Handbook of Human Factors and Ergonomics*, John Wiley & Sons, Inc., 2012, pp. 1016-1030.

[6] SAE Cooperative Research Program, "CAESAR 3-D Anthropometric Database," 13 May 2013. [Online]. Available: http://shapeanalysis.com/CAESAR.htm. [Accessed 25 9 2018].

9 TABLE OF FIGURES

10 AUTHORS BIOGRAPHY

Ken Goh holds the current position of General Manager, at Knud E. Hansen Australia. His responsibilities include advanced concepts and senior mechanical engineering activities. His recent experience includes the concept design of the new Australian icebreaker and a new submarine tender vessel using the ShipSpace™ virtual reality design tool and managing the detailed design of the HVAC system on the Queen Elizabeth Class aircraft carriers.

Robert J Spencer holds the current position of Chief Executive Officer at Stirling Labs. He is responsible for development of the ShipSpace™ collaboration tools. His previous experience includes development of innovative human computer interfaces at Spinfast and successfully conducting the first trial of 1 million simultaneous user avatars in a single shared multi-user virtual environment in cooperation with IBM & BigWorld.

PRESENTATION

TAKING A HUMAN PERSPECTIVE ON MANNING EFFICIENCY AND WORKLOAD ON BOARD CHEMICAL TANKERS

S McLay, Lloyd's Register EMEA, UK

SUMMARY

The Lloyd's Register (LR) Human Factors team is currently engaged in exploring manning efficiency on board chemical tankers. This involves taking a human perspective to delve into the activities on board.

The overall project will analyse data collected from 4 chemical tankers within the same organisation. Each tanker represents different characteristics, including, manning structure, trade routes, and vessel characteristics. This paper describes the work undertaken to date by the Lloyd's Register (LR) team.

The project concentrates on manning efficiency and workload from a human perspective taking into account operational efficiency, ensuring safety on board and effective management of work and rest requirements.

Activities include:

- Review of operating procedures and job descriptions
- Observations of key operations on board
- Interviews with Officers and Crew
- Workload assessment

Some interesting observations and human factors considerations identified during vessel visits are presented. In addition, early conclusions from the project are discussed.

1. INTRODUCTION

There is always a keen interest within the shipping industry to make operations more efficient. The chemical tanker industry recognises that it may be at a critical point in relation to manning numbers. Moreover, it is important to first understand how altering manpower could be achieved whilst preserving the safety of crew and cargo on board.

LR was asked to provide insight to on board activities in relation to manning efficiency across the vessel.

The challenge was to explore how manning changes could enhance operational efficiency whilst ensuring safety. To do this, it was important to look past documented procedures and processes. In fact, to gain knowledge from the vessels themselves as to how operations are conducted in practice. Taking a more detailed approach to identify levels of responsibility on board and the distribution of tasks. A need to establish a picture of communication channels from ship to shore was also identified.

1.1 UNDERSTANDING THE IMPACT OF MANNING CHANGES

There are many proven scientific methods for understanding the manning in a workplace; not many though have been systematically applied to the Marine industry. Using the LR team's cross-industry experience we used a combination of hierarchical and functional task analyses to map current on board activities (shown in Figure 1).

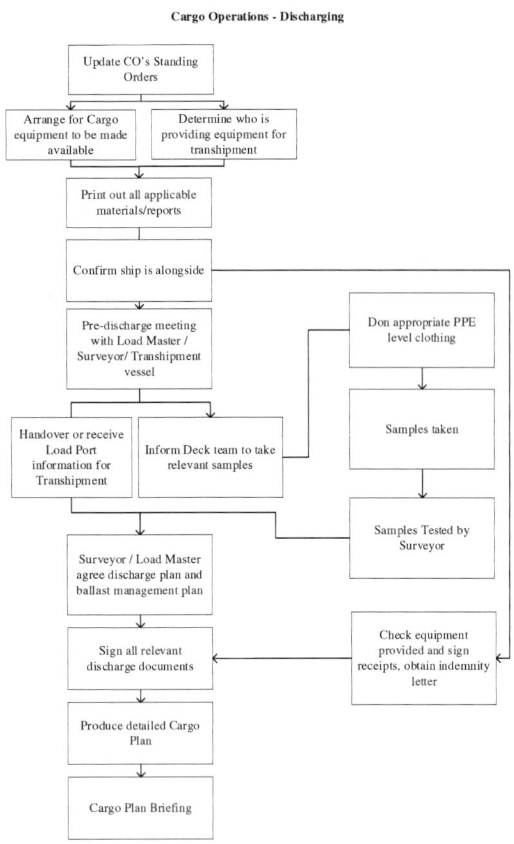

Figure 1 : Example of task analysis for cargo operations

We also reviewed a number of workload assessment tools to identify which was most appropriate given operational and safety constraints on board. Due to the safety-critical nature of Chemical Tankers, we needed a workload scale that would allow the Crew to safely perform their activities, and fill in their workload ratings as accurately as possible.

A hybrid of two recognised workload scales: Integrated Workload Scales (IWS – [2]) and Subjective Workload Assessment Technique (SWAT) [3] were selected. These tools are both self-reported scaled workload measures; IWS (originally from the Rail Industry) assessing overall workload for the period of work utilising a scale of 1 (very low) to 9 (very high), and SWAT assessing more detailed workload for time, mental effort and pressure, each of these factors are scored from 1 (low) to 3 (high).

Observations and semi-structured interviews were also utilised to provide a robustness to the task analyses and further context upon which to identify operational efficiencies.

1.2 AIMS & OBJECTIVES

1.2 (a) Aim

- Determine an optimised on board manning structure for selected shipping routes and vessels.

1.2 (b) Objectives

- Provide guidance to inform sustainable dynamic manning whilst maintaining safe operations. Identify vessel-wide high and low workload and what conditions may contribute to this.

2. METHODOLOGY

2.1 VESSEL DEMOGRAPHIC INFORMATION

2.1 (a) Vessel Characteristics

To date data has been collected from three chemical tankers of vary ages between 2 and 19 years of between 30,000t DWT and 45,000t DWT.

2.1 (b) Personnel

There were a total of 80 personnel across the three vessels visited. Officer nationality varied between vessels with a majority of Filipino crew. Across all three vessels 80% of Officers and Crew were Filipino, 10% Russian. Other nationalities included Latvian, Vietnamese, European, Norwegian and Polish.

2.2 DATA COLLECTION ACTIVITIES

2.2 (a) Desktop Review – Task Analysis

The LR team conducted a review of relevant Standard Operating Procedures and Quality Assurance manuals. They also discussed key operations with subject matter experts from the shore-side organisation.

This information informed the task analyses of key operations to be further refined through information gathered on board the vessels.

The task analyses comprised all key operations on board, including:

- Manoeuvres in port (under Pilotage)
- Navigation at sea
- Cargo operations
- Tank cleaning
- Deck and Engine department maintenance
- Handling stores and provisions
- Bunkering
- Training and Drills

- Port inspections
- Galley operations

On board job descriptions were also provided. This information informed roles and responsibilities associated with the key operations on board.

The expanded RASCI model [1] was used to map the associated roles and responsibilities for each performed activity.

RASCI	Definition
Responsible	Person responsible for the work to be completed; this may be the person who actually does the work or directs other to do the work
Approval	Person who has the legitimate authority to approve the task
Support	Person whose services or coordination is needed
Consulted	People with technical expertise (Subject Matter Expert) whose input adds value for ultimate implementation
Informed	People who need to know / be informed about the status / results of the work, the decisions that were made and who depend on the deliverables

Table 1 : Descriptions of RASCI Methodology ([1])

2.2 (b) Vessel Visits – Observations, Interviews & Self-Reported Workload Measures

LR conducted three vessel visits this year. The organisation identified the target ships based on vessel characteristics (such as age, build and type) and the voyage schedule (e.g. port calls, planned operations, inspections), and manning characteristics. LR sailed with the vessels during the period of data collection.

On board observations were conducted for all major operations occurring at that time. Additional support was provided by a company representative (a Captain) sailing with LR to offer technical clarification as required. This ensured limited distraction or disruption for those directly involved in safety-critical operations.

Individual semi-structured interviews were conducted on board lasting an average of 45 minutes. Interviewees were asked about their role-based tasks and associated workload. Operational conditions which may affect workload and any ideas on how to improve efficiency on board were also explored.

The LR team provided each crew member with a daily workload record sheet. These were to be completed for the duration of the vessel visit.

Crew listed the activities they were involved in each day. Activities were then broken down into tasks. Crew rated their perceived workload for each task they'd been involved in.

Overall workload was scored 1 (very low) to 9 (very high). The lower the rating the more manageable the perceived workload. More detail of workload was acquired through additional self-ratings for time, mental effort and pressure. Each of these scales was scored 1 (low) to 3 (high).

Lower time ratings indicated spare time during that task to complete other tasks. Lower mental effort indicated lower concentration requirement during that task. Lower pressure rating indicated perceived lower stress or demand felt during that task.

The benefit of measuring workload is that it highlights :

- Activities which have high / low workload
- Roles / ranks reporting high / low workload

This data allows conclusions to be drawn regarding distribution of tasks. How a more balanced workload could potentially be achieved. Workload findings are further contextualised by capturing free text comments on the workload sheets. Information reported included conditions which may make operations easier or more difficult. Crew could also use the forms to make suggestions for improvements.

The data gathered directly informs the activity mapping of on board operations and provides detailed information on the workload demands across the vessel. Qualitative information was recorded in a detailed issues register.

2.3 DATA ANALYSIS

Data analysis is ongoing at point of writing. Data collected relating to key operations are currently being explored.

Initial workload analysis is being carried out to identify those roles and responsibilities that experience particularly high and / or low workload demand during operations.

A qualitative data workshop has been carried out to identify any initial patterns in relation to efficiency. These patterns were grouped into themes and input into the issues register for further analysis. Early analysis examines potential impacts and possible operations for efficiency improvements.

3. INTERIM FINDINGS

Data analysis to date has identified some common themes. Remaining data collection and analysis activities aim to explore these and any additional emergent themes. Detailed data validation also needs to take place. Some high-level headlines can, however, be identified at this stage.

3.1 KEY HEADLINES

3.1 (a) Duplication of effort and high volume of paperwork

A high volume of paperwork, both paper-based and digital, to support operations was reported. Some challenges to efficiencies reported include:

- Port / Shore operations requirement for a high volume of paperwork for each operation
- Duplication of signatures on a document reprinted for different departments / agencies.
- Main paperwork burden currently on a few key individuals, e.g. Chief Officer, Chief Engineer.

3.1 (b) Clarity and streamlining of ship and shore responsibilities

Some reported challenges include:

- General perception that shore operations may not fully understand the workload pressures experienced by the crew when performing different operational requirements.
- Ships would like to receive more support from shore-side personnel. Themes included avoiding asking for information already provided, especially during times of high operational demand.
- Some lack in clarity and communication regarding support roles and responsibilities.

3.1 (c) Port stays are often uncertain and difficult to plan

The schedule during port stays is ever changing. Resulting in additional workload and strain on resources across the ships. Although generally accepted as the norm (as it tends to be outside of the Company's control) this can have a significant impact on workload and the ability to plan resources effectively.

Challenges include:

- Additional workload for Officers to perform dynamic planning in-line with multiple schedule changes. Sometimes it is not possible to advance plan.
- Unexpected / unavoidable operational delays due to availability of Port personnel or berths. Leaving crew on standby for long periods of time.
- Miscommunications, conflicting schedule information. The ship is responsible for problem solving with sometimes patchy information.
- Provisions and stores can arrive at a different time to that which is scheduled, and often at short notice making it difficult to plan.
- Resource management to ensure adequately rested and available crew in an ever-changing landscape.

3.1 (d) On-board adaptations to responsibilities are being undertaken at a vessel level to support manning

Examples have been given of local changes on board vessels to support reportedly stretched departments. This entails some re-distribution of tasks as appropriate.

The most common changes could be seen to support the Galley Department. Some Officers were taking responsibility for cleaning their own cabins. This is to support a reduced Galley compliment with their housekeeping duties.

In addition, all available personnel reportedly assist with receiving provisions and stores. This is both time consuming and labour intensive.

Analysis of efficiencies that can be achieved are currently being explored, including: redistribution of personnel responsibilities, different organisational structure and crew complement, review and analysis of systems in relation to exchange and availability of information, addition of novel technology, amongst others.

4. NEXT STEPS

At the time of writing, the following next steps for the project are:

- Data collection on the final vessel
- Further quantitative and qualitative data validation and analysis
- Efficiency analysis and recommendations
- Guidance for optimum manning structure

5. CONCLUSIONS

Despite the project being in progress some common findings have been found. It is recognised that issues identified may have short, medium or longer-term suggestions.

Early conclusions can be drawn about areas which impact manning efficiency on board. These remain areas of focus to take forward to provide suggestions for improvements

6. REFERENCES

1. HIGHTOWER, R., 'Internal controls policies and procedures', 2008.
2. PICKUP, L., WILSON, J. R, NORRIS, B. J., MITCHEL, L., & MORRISROE, G., 'The Integrated Workload Scale (IWS): A new self-report tool to assess railway signaller workload', 2005.
3. REID, B., & NYGREN, T., 'The Subjective Assessment Technique: A Scaling Procedure for Measuring Mental Workload', 1988.

7. AUTHORS BIOGRAPHY

Stephanie McLay MSc, MBPsS holds the current position of Senior Human Factors Consultant at Lloyd's Register. She has 7 years experience as a Human Factors practitioner engaged in project delivery and lead roles across a range of sectors. Her previous experience includes a background in organisational psychology with a particular interest in situational awareness and decision making in safety critical environments. She has been involved in job role evaluation and assessing training and coaching needs within multiple sectors.

LR has Human Factors Specialists with a range of experience in manning efficiency assessments across different industries including the marine and offshore energy market.

AUTHORS' CONTACT DETAILS

Monica Lundh
Chalmers University of Technology
monica.lundh@chalmers.se

Katie Aylward
Chalmers University of Technology
katie.aylward@chalmers.se

Johannes Hüffmeier
Research Institutes of Sweden
johannes.huffmeier@ri.se

Malcolm Cook
BAE System
malcolm.cook2@baesystems.com

Hugo Nguma
Naval Group
hugo.nguma@naval-group.com

Rozenn Coutellier
Naval Group
rozenn.coutellier@naval-group.com

Chantal Mais
Naval Group
chantal.mais@naval-group.com

Viet Dung Vu
Western Norway University of Applied Sciences
Viet.Dung.Vu@hvl.no

Margareta Lützhöft
Western Norway University of Applied Sciences
Margareta.Holtensdotter.Lutzhoft@hvl.no

Linda Sørensen
Western Norway University of Applied Sciences
linda.sorensen@bwfm.com

Kjetil Nordby
The Oslo School of Architecture and Design
kjetilnordby@gmail.com

Synne Frydenberg
The Oslo School of Architecture and Design
Synne.G.Frydenberg@aho.no

Thomas Porathe
Norwegian University of Science and Technology
thomas.porathe@ntnu.no

Sashidaran Komandur
Norwegian University of Science and Technology
sash.kom@gmail.com

Dennis Barber
dennis.barber@talk21.com

Gerd Berner
University of Tasmania
gerd.berner@utas.edu.au

Rory Hopcraft
University of London
Rory.Hopcraft.2014@live.rhul.ac.uk

Jonathan Earthy
Lloyd's Register
Jonathan.Earthy@lr.org

Rafet Kurt
University of Strathclyde
rafet.kurt@strath.ac.uk

Costantino Bongermino
University of Genoa
costantinobongermino@gmail.com

Meyrick Hadfield
Health and Safety Executive
Meyrick.Hadfield@hse.gov.uk

Ken Goh
Knud E Hansen Australia P/L
KEG@knudehansen.com

Joel Scanlan
University of Tasmania
joel.scanlan@utas.edu.au

Beatriz Navas
University of Strathclyde
beatriz.navas-de-maya@strath.ac.uk

Paola Gualeni
University of Genoa
gualeni@dinav.unige.it

Vaughan Pomeroy
University of Southampton
vaughan_pomeroy@hotmail.com

Helle Oltedal
Western Norway University of Applied Sciences
Helle.Oltedal@hvl.no

Stephanie Mclay
Lloyd's Register EMEA
Stephanie.McLay@lr.org